Science Projects
for Young People

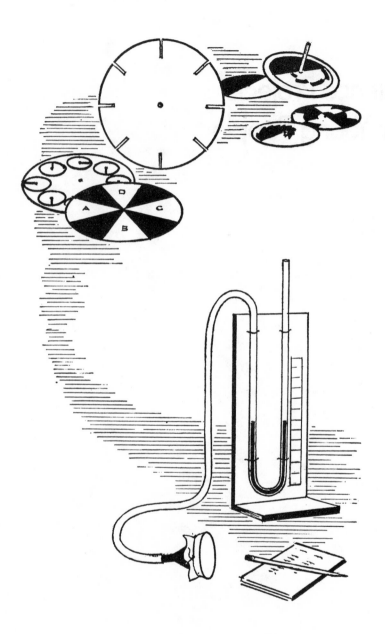

Science Projects
for Young People

by George Barr

illustrated by Mildred Waltrip

Dover Publications, Inc., New York

To Hylda—
who always listens eagerly
and lovingly to my ideas

Published in Canada by General Publishing Company, Ltd., 30 Lesmill Road, Don Mills, Toronto, Ontario.
Published in the United Kingdom by Constable and Company, Ltd., 10 Orange Street, London WC2H 7EG.

This Dover edition, first published in 1986, is an unabridged and unaltered republication of the work first published by the McGraw-Hill Book Company, New York, in 1964 under the title *Research Adventures for Young Scientists*.

Manufactured in the United States of America
Dover Publications, Inc., 31 East 2nd Street, Mineola, N.Y. 11501

Library of Congress Cataloging-in-Publication Data

Barr, George, 1907-
  Science projects for young people.

  Reprint. Originally published: Research adventures for young scientists. New York : McGraw-Hill, 1964.
  Include index.
  Summary: Introduces the methods of scientific research through simple experiments with light and color, sound and music, plants, and chemical analysis. Also discusses equipment and safety.
    1. Science—Experiments—Juvenile literature.   2. Research—Juvenile literature.
  [1. Science—Experiments.   2. Experiments]   I. Waltrip, Mildred, ill.   II. Title.
  Q163.B345   1986   507'.8   86-16825
  ISBN 0-486-25235-3 (pbk.)

Introduction

Do you enjoy discovering answers to original experiments and solving challenging problems? Do you want to test your ability to design simple apparatus and to observe and draw conclusions like a professional scientist? If you are such an adventurous person, then this special book is for you.

Here you will find genuine research problems which can only be solved by your own investigation. Frequently the answers are not like those your friends will get, because their conditions may not be like yours. Sometimes these experiments are never fully completed, because the end of one may only start you off on another problem to solve. This too is how research scientists react all over the world.

You have the ability to do everything in this book. The materials suggested are easily available, safe and inexpensive. At the beginning of each topic you will find sufficient information concerning the concept or principle involved in your research problem. You will also note that many suggestions are given for you to expand your research.

There are many chapters covering a wide range of

topics. This will give you opportunities to work in numerous fields and interests. Here are some problems which you may find particularly challenging:

How can you see the tiny blood vessels in your eyes without using a magnifying glass?

Can you make rapidly turning machines appear to stand still?

How do your eyes help you keep your balance?

How is fresh water obtained by freezing sea water?

What is your horsepower?

Can you make a burglar alarm which is set off automatically when someone picks up your school bag?

How can you see many colors by gazing at a black and white whirling disk?

How far can you see different colors?

Which words are most difficult to hear correctly under noisy conditions?

With simple materials that are easily acquired, you can build a spectacular electric meter. With a potato and a drinking straw, you can find out how a tornado is able to drive a blade of dried grass into the trunk of a tree.

Would you like to learn how to grow real stalactites like those hanging in caves? Will frozen seeds sprout? And how can you detect carbon dioxide in your home?

Many of these experiments are excellent projects for science fairs and clubs. Some will make splendid assembly presentations. Above all, this book will help you learn the care and technique used by research scientists.

Have fun!

4

Contents

Science Projects
for Young People

LIGHT AND COLOR

Can moving things
be made to "stand still"?

Very often engineers and mechanics wish to study the behavior of a swiftly rotating or rapidly vibrating object. In this way they can find out whether a part is under strain or whether it is out of balance. Scientists have discovered a device to "freeze" the motion so it appears to stand still. This useful optical apparatus is called a STROBOSCOPE (STROBE-uh-skope), which in Greek means "whirling view."

You can understand how a stroboscope works by imagining yourself in a dark room, facing an electric fan in operation. If you switch on a light, then, of course, the rotating fan blades will appear as a blur. But suppose that while the blades were turning, you had a way of regularly switching the light on and off, so that the blades were seen in the same position every time they were briefly illuminated. If your switching technique were accurately timed, the fan blades would appear to be standing still.

Most stroboscopes actually work on the idea of illuminating the rotating object with a bright light, which goes on and off at a fast, even rate. This flickering light

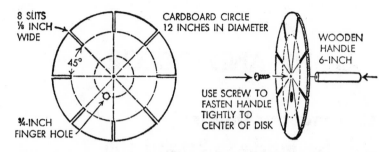

8 SLITS
⅛ INCH
WIDE

7°

45°

¾-INCH
FINGER HOLE

CARDBOARD CIRCLE
12 INCHES IN DIAMETER

WOODEN
HANDLE
6-INCH

USE SCREW TO
FASTEN HANDLE
TIGHTLY TO
CENTER OF DISK

can be carefully regulated by electronic controls. However, you can get almost similar results by making the simple stroboscope shown in the illustration.

It consists of an 8 to 10-inch disk of stiff cardboard, such as the kind used for display purposes in drug store windows. The disk contains equally spaced slits about ⅛ inch wide and about 2 inches long, cut along a radius from the circumference.

The disk is held loosely but steadily in one hand, by means of a ⁵⁄₁₆-inch round wooden dowel. This is glued and nailed to the center as shown. The index (pointing) finger of the other hand fits into a ¾-inch hole, which is about 2 inches from the center of the stroboscope, and turns it at the required rate.

The object is viewed through the slits while you hold your head in one position. It is best to have a strong light on the rotating object, and much less light shining on your side of the disk. Needless to say, the disk must be turned *at a very steady rate*, regardless of the speed of your finger. You will find that if you turn the disk too fast or too slowly, the picture will seem to rotate forward or backward. It takes a little practice to rotate the disk

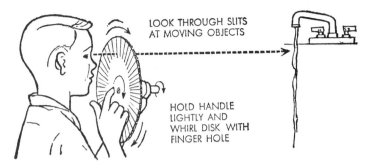

LOOK THROUGH SLITS
AT MOVING OBJECTS

HOLD HANDLE
LIGHTLY AND
WHIRL DISK WITH
FINGER HOLE

by hand at the proper rate, so that the viewed object remains in one position.

For more accurate control you might invent some harmless way to attach the disk to your mother's variable-speed, electric mixer which she uses in the kitchen. Try using string or wire, or you might attach the stroboscope to the shaft of any small motor having a speed control. If you do not have such a control on your motor, the pressure of a cloth or a finger against the shaft can often control the speed.

No matter what you use to turn the disk, make sure that it is perfectly centered and does not wobble. Some experimenters have even attached disks to hand-held, manually operated egg-beaters. Try it. You can get a steady rotation this way.

Try disks with different numbers of *equally spaced* slits. Start with one, then use two, then four, and finally eight slits. Actually, you can make one disk with eight slits, and place masking tape over the slits you are not using. Also, experiment with wider and longer slits, until you get your best results.

Some of the objects you can make appear motionless

11

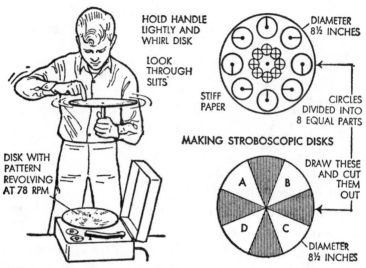

HOLD HANDLE
LIGHTLY AND
WHIRL DISK

LOOK
THROUGH
SLITS

STIFF
PAPER

DIAMETER
8½ INCHES

CIRCLES
DIVIDED INTO
8 EQUAL PARTS

DISK WITH
PATTERN
REVOLVING
AT 78 RPM

MAKING STROBOSCOPIC DISKS

A   B

D   C

DRAW THESE
AND CUT
THEM
OUT

DIAMETER
8½ INCHES

with your homemade stroboscope include: the turning blades of a fan; something rotating on a 78 rpm (revolutions per minute) record player; fast-moving drops of water coming from a faucet, so that they look almost like a continuous stream; the hammer of an electric bell which is ringing; a string with a weight at one end and rotated by hand; any fast-moving parts or gears on machinery.

You can have fun viewing some of the prepared, paper disks shown in the illustration. Copy each carefully on circles of about 8 inches in diameter, and place it on a 78 rpm record player. When using the design with the clock faces, you can make the dial hand turn around and around. The disk with the pie sections can be made to appear to stand still. The letters will help you know whether you are looking at the same section, or a similar section, when everything seems to stand still. Study these disks to see how they were made, and then invent your own interesting designs.

Look through the slits of the rotating stroboscopic disk at its own reflection in a mirror. Why are the slits *absolutely* motionless?

Look at your TV picture through the rotating slits. You will get many designs. Why is this? (Hint: A television screen is constantly being swept by an evenly spaced electronic light beam.)

Buy a stroboscopic, speed-testing disk for your record player, and place it on the turntable. This very inexpensive disk tells you when the speed is exactly 78, 45, or 33⅓ rpm. The proper design stands still when illuminated by a light bulb. Most homes receive 60-cycle alternating current (AC), which make the bulb light up 120 times per second.

Try to design a stroboscope which does not use a rotating disk through which you must look. Instead, you must use a shielded light source such as a box. The box has an opening for the light beam. The disk with the slits is made to rotate in front of it. In this way a flickering light is thrown on the object to be viewed. Movement can be stopped by controlling the speed of the disk. Of course, this is best done in a darkened room.

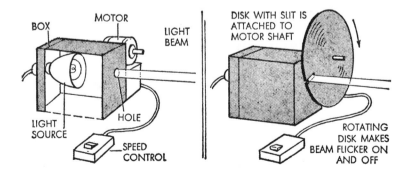

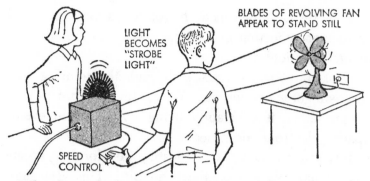

Why do spoked wheels of vehicles often appear to go backward in moving pictures? (Hint: Projectors throw about twenty separate frames per second on the screen.)

Why does a flash of lightning make every moving thing seem to stand still? (Hint: The duration of a lightning flash is often about half a millionth of a second.) Find out about the strobe lights which are used to examine heavy industrial machinery while it is rotating. Ask automobile mechanics how they use stroboscopes to balance wheels while they are turning. Mechanics also use flickering lights to test and adjust the timing of the explosions in the engine cylinders. How are electronic strobe lights used in photography in place of flash bulbs?

How far
can you see colors?

It is very important for many people to know which colors can be seen best from a distance. Highway engineers who design road signs know that some colors stand out much better than others. Sportsmen who go deer hunting wish to wear caps of the most visible color, in order to avoid being mistaken for a distant animal and accidentally shot. Water towers and other high structures are painted in very distinct colors to help airplane pilots see them more clearly.

For greater safety in factories, certain dangerous areas and parts of machines are painted with the most distinguishable colors. When military experts wish to camouflage installations, they are careful to use colors which are not easily detected. Even educators wish to know what colored chalk is most easily seen on differently colored blackboards. No doubt you can think of many more practical uses for the information which you can discover by means of your own experimentation.

If you think about it, you will realize that this problem is not as simple as it sounds. Visibility of a color depends upon many conditions of which you should be aware.

But perhaps before you begin, you should understand why objects have a color. As you probably know, sunlight is composed of all the colors of the rainbow blended together to produce a white light. These colors are roughly red, orange, yellow, green, blue, and violet.

When sunlight shines on different substances, not all the colors are reflected equally. We say that an object has a certain color because of the colored light it reflects to our eyes. All the other colors in the sunlight mixture, which are not reflected, are absorbed by the object.

For example, grass is green because it reflects mainly the green color of the sunlight. All the rest of the light is absorbed by the grass. In the same way, a banana skin reflects yellow and absorbs all the other colors in the sun-

15

light. Other substances reflect several colors at one time, the combination of which gives us many color variations. For example, if only red and blue are reflected, then the resulting color would be purple. If all the colors are reflected, a substance appears white. But if no light is reflected, then an object is black.

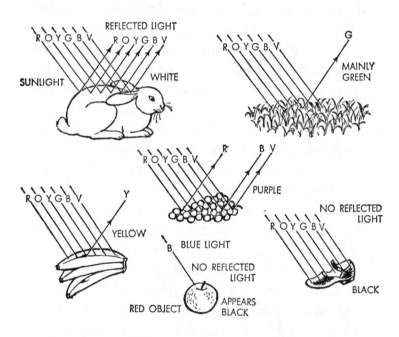

When doing color experiments, you must remember that the color of something depends upon the light which is shining on it. A red object appears red, only when viewed in white or red light. If the light shining on it has no red in it, then it cannot reflect red. So if only blue light shines on a red object, it would appear black.

When the sun is low in the sky, because of atmospheric

conditions, there is more red in the sunlight. Therefore everything may look redder, and all the color values may change.

To start your experiments, first obtain many colored materials and cut them to identical sizes. Design some hanger for them, so that the colors can be hung in orderly arrangement, from left to right. Have a friend place them so far away from you that you cannot detect the colors. You may find that this distance is too far for comfort. But if you make the colored materials smaller, then the distance will be lessened. As you, or any other observer, walk toward the materials, record with a pencil and paper the color you see first. Also in which place on the hanger you see it. Which do you detect next? Record that you saw it second and also its place on the hanger. Which is the last to be recognized?

Change the positions of the colors on the hanger for each observer. It is not good technique if one knows in advance where each color is placed. After every test, check whether the observers referred to the colors correctly, as they were placed on the hanger. If they were wrong, then you should not use their observations.

Repeat the experiments using different kinds of materials, such as cloth, cardboard, fluorescent paint, glass, etc. Try different light conditions; early morning, noon, sunset, ordinary light bulbs at night, or fluorescent and mercury-vapor street lights. Does it make any difference when there is cloudiness, rain, snow, or fog? Which color stands out best against different colored backgrounds?

Experiment with different people, and see if you get the same results, since no two people see colors exactly alike. It is a fact that 8 males in 100 are color-blind to one color or another. Wouldn't it be surprising if you found out during this experiment that you were somewhat color-blind yourself?

Does it matter if you look from the side of your eyes? Experts say we see less light but are more sensitive to color this way.

When recording your results indicate all the conditions carefully. In this way you can make accurate comparisons. Try to improve the sample below, or copy this one in your notebook.

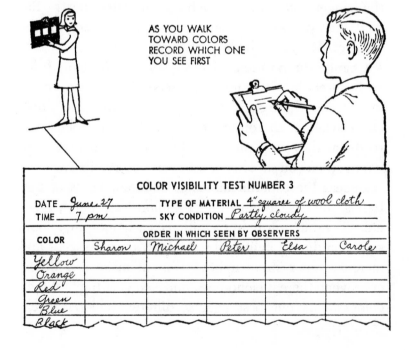

AS YOU WALK TOWARD COLORS RECORD WHICH ONE YOU SEE FIRST

COLOR VISIBILITY TEST NUMBER 3

DATE _June 27_   TYPE OF MATERIAL _4" squares of wool cloth_
TIME _7 pm_   SKY CONDITION _Partly cloudy_

COLOR	ORDER IN WHICH SEEN BY OBSERVERS				
	Sharon	Michael	Peter	Elsa	Carole
Yellow					
Orange					
Red					
Green					
Blue					
Black					

What colors are the road signs in your neighborhood? Do the colors agree with the results of your own research?

When are colors produced from black and white?

Once on a science TV program, somebody twirled the black and white disks shown on page 21. The viewers at home were amazed to see the sections of circles seem to close and form *colored* rings on their black and white sets! The outer rings differed in color from the inner rings. However, when the disks were spun in the opposite direction, the colors of the rings often surprisingly changed places.

You can have many hours of research by making and spinning these disks. The colors which form will depend upon the speed of rotation, the size of the broken sections, and how much of the circle is solidly blackened. You will also discover different colors when the disks are viewed by sunlight, and by artificial light. You may, no doubt, find other reasons for observed color variations. As you know, much depends upon the condition of the observer's eyes too.

Scientists have known about this effect for over 100 years. But no one has yet come up with a good reason for it which will satisfy everyone. Most investigators agree that fatigue of the color-receiving part of the retina has something to do with it. It is also a well-known fact that impressions of different colors remain on the retina

19

for different lengths of time, after they are no longer seen by the eye.

One explanation may be that it takes different lengths of time for different color impressions to be formed in the eyes and to be interpreted by the brain. For example, blue goes through this process more slowly than red.

Here is how the person gets the illusion of color from black and white. As you know, the white portion of the disk reflects all the colors. But each spinning white space is seen by the eye for a very short interval, because it is followed by a black part.

By varying the lengths of black and white, the eye and brain will record only certain colors. This will explain why the colors change as the disks slow down, since there is more time for the eye to see the white part. Of course, this theory does not explain why the colors are reversed during an opposite rotation, since the lengths of the black arcs are the same.

Make the disks of stiff cardboard, 4 inches in diameter. Use black India ink, or a black "magic marker." At first make the arcs ¼ inch thick. Later you can vary the thickness to see if this changes the effect.

Put a pin through the exact center. Hold the sharp end of the pin in one hand, and strike the edge of the disk with the other hand. The disk should rotate with hardly any wobble. Look at the colors. See them change as the disk slows down. It may take a little practice, but in a very short time you will become an expert at reporting your observations. Try it in sunlight, fluorescent light, and also under ordinary light bulbs.

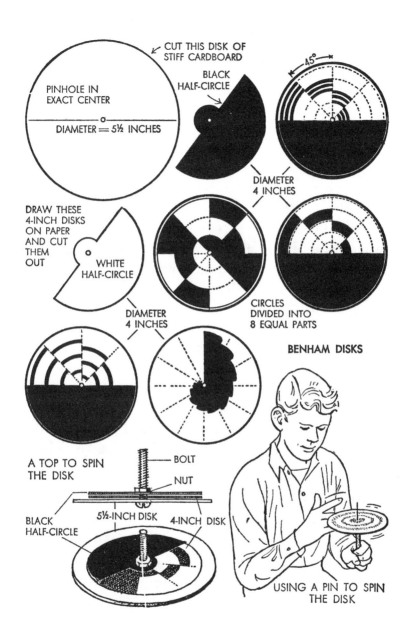

CUT THIS DISK OF STIFF CARDBOARD

PINHOLE IN EXACT CENTER

DIAMETER = 5½ INCHES

BLACK HALF-CIRCLE

45°

DIAMETER 4 INCHES

DRAW THESE 4-INCH DISKS ON PAPER AND CUT THEM OUT

WHITE HALF-CIRCLE

DIAMETER 4 INCHES

CIRCLES DIVIDED INTO 8 EQUAL PARTS

BENHAM DISKS

A TOP TO SPIN THE DISK

BOLT

NUT

BLACK HALF-CIRCLE

5½-INCH DISK

4-INCH DISK

USING A PIN TO SPIN THE DISK

Another way of spinning a disk is to make a top by inserting through the center of the cardboard a small bolt with the head down. Screw a nut on the bolt tightly against the cardboard. Spin the bolt between the fingers as you let the top go. Another kind of top uses a lollypop stick inserted partway through the center and wedged or glued to the disk.

Make separate disks, each time varying the amount of solid black. Better still, if you make a blackened half circle as illustrated, you can vary the black section by placing it in different positions on top of another disk. Also make another half disk of pure white and experiment with that. What is the effect of using colored arcs or colored half sections?

Physicists call the device with which you are experimenting a *Benham* disk. Can you discover other ways of turning the disks; toy musical tops, toy motors, electric food mixers? Probably the best method is to use a hand drill or a variable-speed electric drill. Insert the center bolt of the previously described top into the chuck of the drill. You may use a flat-headed nail instead of a bolt.

Can a pendulum fool you?

You can have many hours of enjoyable experimentation with one of the most amazing optical illusions. Set a pendulum swinging in a straight line. When you look at it through a dark glass held over one eye, the pendulum will appear to move in a circle!

Construct the pendulum by fastening a small ball or other weight to the end of an 18-inch length of light string. You can have a friend hold the top of the pendulum while it swings. Or you can attach the string to the end of a ruler or stick, held out from a shelf or a table, with its end supported by some means. When you release the pendulum weight, make it swing in a straight line exactly left and right, as you will see it from a position which you will take across the room. (In other words, the pendulum should *not* swing toward you and away from you.)

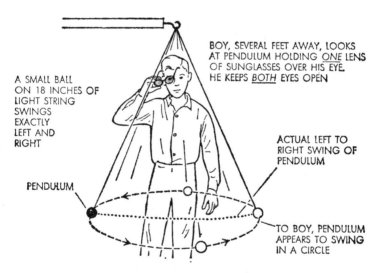

A SMALL BALL ON 18 INCHES OF LIGHT STRING SWINGS EXACTLY LEFT AND RIGHT

PENDULUM

BOY, SEVERAL FEET AWAY, LOOKS AT PENDULUM HOLDING <u>ONE</u> LENS OF SUNGLASSES OVER HIS EYE. HE KEEPS <u>BOTH</u> EYES OPEN

ACTUAL LEFT TO RIGHT SWING OF PENDULUM

TO BOY, PENDULUM APPEARS TO SWING IN A CIRCLE

Now look at the swinging pendulum while holding one lens of your sunglasses over your right eye. Your left eye should be clear. *Keep both eyes open* while looking at the moving pendulum.

You will be delightfully surprised to discover that the pendulum is now not moving left and right in a straight line, but it seems to be swinging in a circle! You will also see something more astonishing, if you look at the pendulum with one lens of the dark glasses over the other eye. This time the pendulum will appear to be moving in a circle in the opposite direction! Remember to keep both eyes open.

Scientists do not know the exact reason for this illusion. Some psychologists think that part of the explanation might be that the eye which is darkened receives the impression, and sends the message to the brain a little slower than the brighter eye. The illusion is created when the brain combines the images.

Does distance make a difference in this remarkable illusion? Will all people see the pendulum swing in the same direction when the same eye is darkened? Try many variations using darker glasses or blackened photography film. How about different sizes of string? Experiment with various light and dark backgrounds. Would there be a difference when viewed against wallpaper having a vertical, horizontal, or other design?

View the pendulum in bright sunlight and also in electric light. Also try it in a dimly lighted room, and shine a flashlight on the pendulum whenever it is in a certain position. See the effect when looking through colored cellophane held over one eye, or even putting different colors over both eyes.

If you experiment with dozens of different conditions, you may discover facts which will help you make more intelligent guesses about the explanation of this curious phenomenon.

ELECTRICITY AND MAGNETISM

Can you design
an electric meter?

Would you like to construct an electrical instrument that will measure the flow of electricity? The heart of this device is a kind of electromagnet called a SOLENOID (SOLE-en-oid). It consists of many turns of insulated wire coiled around a hollow tube.

When the bared ends of this coil are connected to a source of electricity, the coil becomes an electromagnet. A rod of iron, which is placed partly into one end of the tube, will be drawn into the tube of the solenoid. When the circuit is broken, the solenoid loses its magnetism. The iron plunger is drawn out of the tube by means of a weak spring, rubber band, or by the force of gravity as in our case.

See the illustration on page 29. The solenoid is activated by being connected to a dry cell. It draws in the large nail. This pulls down the small end of a large, pivoted pointer. Notice that this arrangement allows the *small* movement of the nail to make a *large* movement of the end of the pointer. The weight of the large pointer is also used to pull the plunger out of the solenoid when the circuit is broken. It works like a lever.

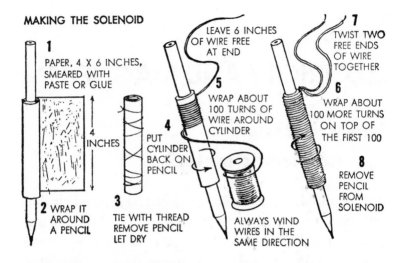

1 PAPER, 4 X 6 INCHES, SMEARED WITH PASTE OR GLUE

4 INCHES

2 WRAP IT AROUND A PENCIL

3 TIE WITH THREAD REMOVE PENCIL LET DRY

4 PUT CYLINDER BACK ON PENCIL

LEAVE 6 INCHES OF WIRE FREE AT END

5 WRAP ABOUT 100 TURNS OF WIRE AROUND CYLINDER

ALWAYS WIND WIRES IN THE SAME DIRECTION

7 TWIST TWO FREE ENDS OF WIRE TOGETHER

6 WRAP ABOUT 100 MORE TURNS ON TOP OF THE FIRST 100

8 REMOVE PENCIL FROM SOLENOID

The best wire to use for winding the solenoid is no. 22 cotton-covered copper wire. This comes in very inexpensive spools. You will need only about 23 feet for a 200-turn solenoid. Ask a radio repairman or a hardware dealer to get you some.

This wire does not have thick insulation. In this way the turns can be closer together, and the magnetism is concentrated. Most commercial electromagnets are wound with copper wire, which is insulated with only a thin coat of enamel. You can use this too, but you will have to handle the wire more carefully than the cotton-covered kind. You need never buy wire if you wish to take apart broken bells and other discarded electrical devices using these wires. In a pinch you may even use any thin bell wire you may have. If necessary, splice several lengths.

The turns of wire must be wrapped about a stiff tube of nonmagnetic material. It should be 6 inches long and about the diameter of a pencil. If you cannot find such a tube, you can make it as follows:

Get a 4-inch by 6-inch rectangle of rather heavy paper, such as drawing paper. If you use ordinary writing paper, make the rectangle 4 inches by 8 inches. Place it flat and smear the entire top surface with mucilage, library paste, or a mixture of flour and water. Roll this around a long pencil into a tight, even, 4-inch tube. Wrap some string over the entire length to prevent unwinding.

Dry the glued tube in a warm place if possible. Keep twisting the pencil inside the tube, so that the glue does not stick to the pencil. After a while remove the pencil, and allow the tube to stiffen as it dries thoroughly. Remove the string too.

When ready to wind the wire, put the tube over a pencil again for support. Leave about 6 inches of wire free for making connections. Make your first loop around the tube about ¼ inch from the left end. Twist the wire tightly on itself several times, so that the end wires do not unravel as you work. Wind evenly. About every ten turns twist the pencil, so it is not clinched permanently inside the tube.

Wind until you come to about ¼ inch of the right end of the tube. Make your second layer of turns back over the first layer, going from right to left. However, do not reverse your direction of winding each turn. All turns on a solenoid must be wound in the same direction. When you reach the starting point of the first layer, you must prevent the coil from unraveling. Do this by carefully twisting the wire you are winding around the wire which started your coil and is extending out. Do not damage the insulation of either wire. The two wires must not make bare-metal contact. Put some adhesive tape around one wire, before you twist the two wires. Perhaps you can finish off your coil in a different, neater way to prevent unraveling of the turns.

The solenoid's plunger is a 4-inch steel "20-penny common nail" used by every carpenter. The "20-penny" refers to an old English weighing system and not to the price! Try to get a nail that does not have deep ridges below the head. If it does, use a file or sandpaper to smooth the grooves down, so the nail will have no trouble entering the solenoid. The string is attached to the nail, as shown in the illustration, in order for the nail to be suspended vertically. It is also tied to the pointer in such a way that it can be moved to change the leverage. This is very important and delicate.

The backboard for your electric meter should be a 12-inch square piece of plywood or other wood. The wooden base should be wide enough to prevent toppling when the back is kept vertical. Place the solenoid about 1½ inches from the end. Keep it in place by wiring it to the backboard, or by means of cloth or staples made of heavy paper placed across it and tacked down. Wrap the ends of the solenoid wires around the connection

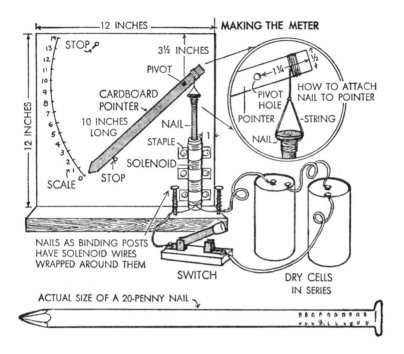

12 INCHES

3½ INCHES

STOP

PIVOT

CARDBOARD POINTER

10 INCHES LONG

NAIL

STAPLE

SOLENOID

STOP

SCALE

12 INCHES

12 INCHES

1¼

½

PIVOT HOLE

POINTER

NAIL

STRING

HOW TO ATTACH NAIL TO POINTER

NAILS AS BINDING POSTS HAVE SOLENOID WIRES WRAPPED AROUND THEM

SWITCH

DRY CELLS IN SERIES

ACTUAL SIZE OF A 20-PENNY NAIL

nails as shown, removing the insulation first.   Binding
posts, of course, make better connections.

The pointer should be a very lightweight strip of
wood.   The weak force of the solenoid must be able to
lift the pointer as the nail is drawn down.   At the same
time, the length of the pointer should enable it to balance
the nail when it is *out* of the solenoid.   It is simply a
matter of leverage.   Moving the suspended nail *away*
from the pivot brings the pointer up.   Needless to say,
the pivot should be loose.   Use a washer at the pivot to
keep the pointer from scraping the backboard.

29

Start with the dimensions illustrated, then modify slightly so that the meter works most sensitively with your own materials. Connect your solenoid to a large 1½-volt dry cell. Then use several cells hooked up in series, as shown. Also use a flashlight cell. You should get different readings on your scale.

The scale does not tell you the exact number of volts or amperes. Your toy meter is not designed for such accuracy. Nevertheless, you can get excellent comparison readings for different sources of electricity. It will also indicate that there is a flow of electricity in a circuit.

You can greatly increase the sensitivity by varying many conditions, such as: changing the number of turns on the solenoid, also the size; having the nail at different depths in the tube when taking a reading; using many different types of pointers, and methods of lifting the nail back to the zero position.

Read about solenoids in encyclopedias and science books. They are very useful in industry as well as in your own home. For example, in your washing machine there are many solenoids. Some open and close hot and cold water valves; others engage or release clutches and gears as the washer goes through its cycles.

Can you burglar-proof your school bag?

Do you want to have a barrel of fun at your school's science fair? Invent some tricky burglar alarm for a valise, school bag, large lady's pocketbook, or other carrying case. Your sign might say:

When the bag is picked up, a mighty racket comes from within the bag, much to the puzzlement and entertainment of your classmates. The noise can be made by a loud electric bell, a bicycle siren, or anything else which can be operated by dry cells.

Of course, there is no limit to the ways in which this kind of alarm can be made. It depends only upon your imagination and ability to work with your hands. The simple method described here gives the basic circuit for any system. It consists of a dry cell, bell, and switch.

The bottom and backboard shown in the illustration should be made of fairly heavy scrap wood. The size of the base should be large enough to fit tightly into the bottom of your bag. The alarm switch is made by bending

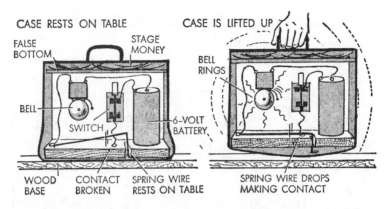

CASE RESTS ON TABLE

FALSE BOTTOM

STAGE MONEY

BELL RINGS

BELL

SWITCH

6-VOLT BATTERY

CASE IS LIFTED UP

WOOD BASE | CONTACT BROKEN | SPRING WIRE RESTS ON TABLE

SPRING WIRE DROPS MAKING CONTACT

springy clothes-hanger wire with pliers. Use steel wool or sandpaper to scrape off the black enamel wherever electrical contacts are made.

The slit in the wooden base allows the switch to be brought toward the center. If you wish, you can drill a hole for the wire instead of sawing the slit.

The wire-hanger switch should be tight against the nail-head contact. If it is loose, put a pencil under the wire near the nailed-down end. Now bend down the free end of the wire, close to the pencil. Remove the pencil and the spring snaps down tightly.

The bottom loop on the wire may go through a hole in the bottom of the bag. However, most school bags have a soft bottom and tend to sag. Test yours. All you need is a very small movement every time the bag is lifted or set down. If yours has this looseness, it will not be necessary to make a hole. This adds to the mystery. At any rate, try to devise some system where no part of the wire will be visible.

The knife switch is really not necessary. But it is good

to have a cut-off switch, just in case the bag falls over at some time, or has to rest on its side.

Build a cover over the unit, which will act as a false bottom, and "fill" the bag with stage money. To make more of a clatter, try connecting a 6-volt dry battery to the bell, siren, or horn. Can you invent a system where only you will be able to pick up the bag without sounding the alarm? Use some secret switch.

Design an alarm which is not contained in the bag. Instead, some other alarm goes off when the bag is removed from some box or platform. Can you make one without a piece of wire or a switch protruding from such a platform? (Hint: Place a strong magnet in your bag. When it is set down over, and close to an easily moved, iron switch under the platform, the circuit is broken.)

How about a real show-stopper using the following idea? You may have a small battery-operated tape recorder, which is so popular today. Fix up a valise so that when somebody tries to walk off with it, a loud voice inside screams, "Stop thief! Help! I'm being stolen!"

Does a magnet get weaker when it rusts?

Since most magnets are made of iron, they will rust when they are kept in moist places. You have often seen some of these. Do you think that a rusty magnet is not as good as when it was polished and shining?

A magnet consists of an orderly arrangement of most

of its molecules. Magnets can be spoiled by doing something which will make these molecules move out of the magnetic pattern. For example, banging or repeatedly dropping a magnet, will jar the molecules so that they will become disarranged.

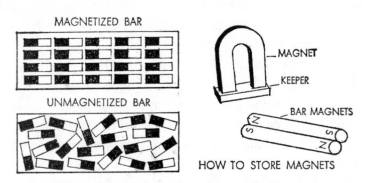

MAGNETIZED BAR

UNMAGNETIZED BAR

MAGNET

KEEPER

BAR MAGNETS

HOW TO STORE MAGNETS

Heating a magnetized needle red-hot will also ruin the magnetism. As you know, molecules of all substances are constantly in motion. Normally, they move within certain limits. But when heated, the molecules move faster. There is more chance for molecules to get out of some orderly design into which they were forced to go when they were magnetized.

Placing a magnet next to other magnets can also force molecules out of position. It is well known that when magnets are stored they should have a bar of iron placed across the poles. This bar is called a KEEPER. Without keepers different magnetic fields would act upon each other to weaken them.

These effects, which weaken magnets, occur throughout the metal. Rusting, however, takes place only on the surface. The total amount of iron combining with oxygen to form rust is very small. Of course, if a bar of steel were to rust for many years the story would be different.

Test how much strength a magnet loses when it is rusted. First find out how strong a shiny new magnet is by counting how many small, similar-sized nails it will pick up. Always dip the magnet into the nails in the same way. Do it many times and get an accurate average figure.

Place the magnet into a jar containing some water. Keep it there until it shows a rusted surface. Can you think of other ways to hasten the rusting? Will salt water corrode iron faster? Is a moist atmosphere better for rusting than actually immersing the iron in water?

Make sure there is nothing near the jar which might influence your test. Keep other magnets away. Do not even set the jar on an iron surface. Electric motors and other devices may have magnetic fields too.

Test the strength of the rusted magnet in the identical way you tested it before it rusted. Is there much difference?

It might be interesting to have the magnet weighed on the delicate scale of a friendly pharmacist or chemist. Do this before and after the experiment. Then you could know how much (if any) of the magnet was actually lost by the rusting. Compare this with the loss of magnetism (if any).

Rust is iron oxide. It weighs more than the iron itself. Do you think that if the rust does not scale off, you may

RECORD OF STRENGTH OF MAGNETS								
NUMBER OF THE MAGNET	NUMBER OF NAILS PICKED UP							
	9/63	2/64	6/64	7/64				
4	21	17	15	13				
3	14	14	11	8				
7	27	21	6	4				
2	29	28	28	29				

even find that the rusted magnet weighs more than before?

Does every magnet lose some of its strength in time? Keep accurate records of the strengths of all your magnets. Use the same small nails in all your tests. It should be interesting to number all the magnets used by your science teacher. Have a card for each one. Keep a periodic log of its loss of strength. Make a chart or graph of some of these. Plot time in weeks or months, against number of nails picked up. What do you think is weakening the magnets?

WATER AND EARTH SCIENCE

How accurately can you measure water pressure?

Scientists are usually surrounded by expensive, delicate instruments to help them in their research. But for a very small sum you too can build a sensitive and accurate meter to launch you on an exciting round of explorations. It is called a MANOMETER (muh-NOM-iter), and it is used to measure differences in water and air pressures. It can be used for many experiments.

You will need about 6 feet of clear plastic tubing, the kind you can see through. The opening should be about ¼ inch. You can buy this in pet or hobby shops, since it is used for connecting aerators and filters in fish tanks. Druggists and some hardware dealers also carry it.

Make the stand of any wood you have available. The base should be heavy or wide so that the backboard is not top-heavy. An excellent backboard is a piece of ¼-inch plywood, 18 inches tall and 8 inches wide. Nail the plywood to the base.

Attach the tubing to the backboard in the form of a large U as shown in the illustration. Use iron staples large enough not to pinch the tubing too much. You

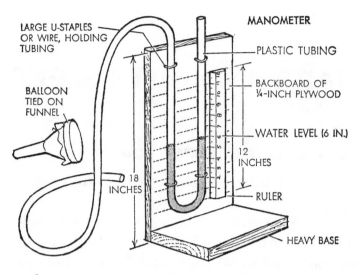

LARGE U-STAPLES OR WIRE, HOLDING TUBING

BALLOON TIED ON FUNNEL

18 INCHES

MANOMETER

PLASTIC TUBING

BACKBOARD OF ¼-INCH PLYWOOD

WATER LEVEL (6 IN.)

12 INCHES

RULER

HEAVY BASE

may also use some wire to hold the tubing to the board. Make small holes through the wood, draw the wire through, and twist the ends together.

The long end of the tube may remain in one piece, leading to the left side of the manometer on the backboard. Or if you wish, you may cut it close to the board. Find a connecting tube of plastic or brass to fit snugly into each cut tube. Tropical fish stores sell these, as well as valves and other fittings.

The dimensions given above can be varied. Even the plastic tubing is not necessary, if you can get someone to bend some glass tubing into the proper manometer shape. Remember, however, that plastic will not break. Rubber tubing may also be used, except of course, for the manometer itself. Attach a ruler vertically and close to the tube of the manometer. If you wish, you can make your own equally spaced lines on the backboard. Perhaps you can invent a way of moving the attached ruler, so that its zero end can always be placed at the water level of the manometer.

Take the manometer near a sink or washtub and pour water into the vertical end of the tube. Stop when you have about 6 inches in each side of the U. Some experimenters put red ink into the water. This may make it more visible from a distance and very impressive, but it also stains things badly. So start your experimentation with water as the indicator.

Notice that the level is the same in each side of the manometer. That is because a manometer measures differences in air pressure. Since both ends are open to the *same* air, the pressures are equal. Now blow *gently* into the long end of the tube. See how the water column on the right side goes up. Your breath has increased the air pressure on the left side. It is now greater than the outside air pressure.

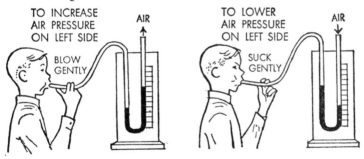

Now suck in some air very gently. The left level rises as the right level goes down. The outside air pressure is greater than the air pressure over the left side. Observe how sensitive this instrument is. For higher pres-

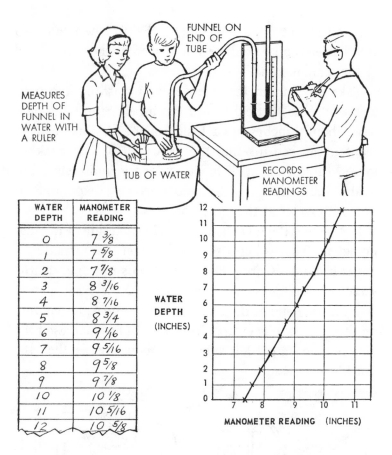

WATER DEPTH	MANOMETER READING
0	7 3/8
1	7 5/8
2	7 7/8
3	8 3/16
4	8 7/16
5	8 3/4
6	9 1/16
7	9 5/16
8	9 5/8
9	9 7/8
10	10 1/8
11	10 5/16
12	10 5/8

sures the right side may be closed by bending over the tube and tying it in the folded condition. A pencil may also act as a plug. Of course, a tight valve placed at the end is excellent. Try blowing hard into the tube, when the opposite end is closed. See the smaller movement of the water. You can even use this device to compare your friend's blowing power.

Scientists also use a heavier liquid than water. Perhaps you have seen manometers containing mercury. Your doctor uses a mercury manometer for measuring blood pressure. Ask him how it operates. Read about it in encyclopedias or physics textbooks. It is called a SPHYGMOMANOMETER (SFIGMOH-muh-NOM-iter).

Use your manometer now for an experiment to find out how water pressure changes with depth. Fill a pail, washtub, or washing machine with cold water. Stand your manometer close by. Grasp the end of the long tube and lower it *vertically* into the water. Do not squeeze it.

See how rapidly the water pressure increases as you lower the tube. Is the increase even? That is, for every inch of depth, is there an equal increase in pressure? For greater accuracy, attach your tube to a ruler by means of rubber bands. Simply lower into the water until the ruler shows the depth. Make a chart and graph like the one shown. Of course, your figures will be different.

Will the pressure be the same for very warm water as it will for cold water? How great is the pressure at the bottom of a swimming pool or a pond? You may have to use a closed-end manometer. Experiment with many different ideas dealing with water or air pressure.

The tube must be held *vertically* or air will escape from the end. This will cause faulty readings. To avoid this escape of air, and for greater sensitivity and accuracy, place a small toy funnel at the end of the tube. Stretch a sheet of thin rubber (obtained from a large balloon) over the funnel. Hold the rubber tightly by means of string or rubber bands, as illustrated. This is done best

41

on a plastic or aluminum funnel having a short vertical section, before it becomes the typical slanting shape of a funnel.

The water levels in the manometer will not be equal now. Hold the rubber-covered funnel in the air, right side up, upside down, left and right. The manometer levels should not change. This shows that air pressure is the same in all directions. Repeat this under water. Water pressure, too, at any depth, acts the same in all directions.

Water is fairly heavy, as you realize whenever you have to carry a pail of water. One cubic foot of fresh water weighs about 62½ pounds. That means that 1 foot below the surface of the water, the pressure on each square foot is 62½ pounds. Two feet below the surface, the pressure is 125 (62½ × 2) pounds per square foot, etc.

Because water pressure increases so rapidly, divers and submarines cannot go down very deep. However, a special diving ship called a BATHYSCAPHE (BATH-ih-scaife) can go down about 13,000 feet. A cubic foot of salt water weighs 64 pounds. The water pressure at 13,000 feet is about 5,900 pounds per square inch. That means that every square foot of the bathyscaphe has about ¾ of a million pounds of pressure acting on it. No wonder the steel walls are thicker than your fist!

Dams are built thicker on the bottom because the water pressure is much greater there. Another interesting and quick way to show the increase of water pressure with depth is to use a quart-size paper milk container. With a small nail, punch a hole horizontally near the bottom. Make a similar hole about 2 inches from the top, and another hole halfway.

Fill the container with water from the faucet. Quickly place the container on an inverted glass standing in the sink. You will see a long, high-pressure stream coming from the bottom hole. The water will come out with less force from the middle hole. The stream from the uppermost hole will show least pressure. The stream will shoot out least.

Make taller "water towers" from two or three milk containers glued into one long one. Or use other taller containers for this demonstration. Can you see now why water towers in your community are able to increase water pressure? Look for them.

Can sea water be desalted by freezing?

One of the biggest problems facing almost every country today is how to get enough fresh water for its needs. This is such a universal difficulty because the population is increasing everywhere. People require more water for drinking, washing, and for thousands of industrial uses, including irrigation.

In this country, almost every one of our states is experiencing a water shortage. Even our way of living makes us need more water. Daily baths and showers consume a tremendous amount of precious water. So do washing machines for clothes and dishes, swimming pools, and air conditioning. We wash our cars more and also water the lawns of more suburban homes. Almost every product we buy uses huge volumes of water in the manufacturing process.

We must conserve water by using it more wisely. We must stop polluting our existing water supplies. Above all, we must find new sources. Many scientists believe that our best hope is to develop ways of getting fresh water from the ocean. More than three quarters of the earth is covered by sea water. What a blessing for the entire world, if a cheap way could be found to desalt ocean water!

In our country, the Department of the Interior has established the Office of Saline Water. Here, scientists are constantly working on inexpensive methods of getting fresh water from the sea. In other countries too, especially in Israel, much similar research work is being done.

There are two main ways of desalting ocean water. You can either take out the water, or else take out the salt. You are probably familiar with the first method. It is a process called distillation, and it removes water from a salt solution by heat. The heating can be done by electricity, fuels, or by the sun's rays. The water boils or evaporates off. The water vapor then condenses back to fresh water as it cools, and the salt remains behind. However, this process is too costly today to produce great volumes of fresh water.

Experimenters have several ways of taking salt *out* of the water. One of the most promising methods is to freeze the salt water. It is based upon the fact that when salt water is frozen, the ice which is formed is practically pure water.

In one of these processes, a volume of sea water is partially frozen. The ice floats on the top and is removed. This ice is now washed in fresh water, and then remelted. It may be frozen again, taken off the top, washed, and remelted. It may be necessary to repeat the entire procedure. The water eventually becomes fresh enough to drink.

Of course, this brief summary is a greatly simplified version of a complicated, technical operation. The method works, but at present it is still too expensive to be very useful.

You can see for yourself how this desalting is possible by doing the following experiments. First, make up a salt solution by dissolving 1 tablespoon of table salt in 1 quart of cold tap water. Stir thoroughly.

Save about half a glass of this solution in some covered container. It is your CONTROL which you will need for comparison purposes. Later, you will compare the amount of salt in a similar volume of melted "desalted" ice.

Pour some of this salt solution you prepared into a tall drinking glass. Since water expands when it freezes, do not fill or cover the glass. Place it into the freezing compartment of your refrigerator. If your folks get worried, tell them the open glass will not crack. About every half hour take a peek at the liquid. Do not allow it to freeze solid.

When the water is about half frozen, remove the con-

1 PUT 1 TABLESPOON OF TABLE SALT INTO 1 QUART OF WATER

2 POUR HALF OF SALT WATER INTO A GLASS

3 POUR OTHER HALF OF SALT WATER INTO A SECOND CONTAINER

4 PUT GLASS INTO FREEZER UNTIL WATER IS PARTIALLY FROZEN

CONTROL

5 STRAIN INTO A CHEESECLOTH

EQUAL VOLUMES

MELTED ICE

CONTROL

6 DIP QUICKLY IN ICE WATER SEVERAL TIMES TO RINSE OFF SALT

7 HEAT BOTH MELTED ICE AND CONTROL WATER VERY SLOWLY

8 COMPARE AMOUNT OF SALT LEFT IN EACH PAN

tainer. You will probably find that the ice has not frozen into a solid block of ice. Instead, the ice is in a slushy form. Remove the ice with a spoon and place it into a strainer made of cheese cloth, clean gauze, or an old silk stocking.

Quickly squeeze out the excess salt water. Now dip the enclosed ice several times into a pan of tap water containing many ice cubes. The chilled water prevents loss of your ice by melting while you are washing the salt from the ice. Finally, rinse the ice in the cheese cloth a few more times in another pan of ice water. Place your ball of strained and washed ice into a clean glass. The entire washing operation must be done fast because the slushy ice is melting fast.

You can now wait until the ice melts. But if you are impatient, you may put all the remaining ice into a pan and melt it by heating *for a few seconds* on the stove. Taste this solution, and also your original control solution, by dipping a clean finger into each. Rinse your mouth between each tasting test. Both liquids should also have about the same temperature. Can you detect any difference?

If you cannot, it may be because you could not wash all the salt water from the mushy ice. When this process is done commercially, the solid ice which is formed is washed thoroughly.

However, you can compare the amount of salt in each container by using a method not involving tasting. First measure how much melted ice you finally have left.

Place this volume into an aluminum fry pan. Measure an *equal volume* from your control container. Pour this into a separate aluminum pan, preferably of the same size. Identify the pans in some way so you do not confuse them. Evaporate the water from each. Compare the amounts of salt which remain.

The heating can be done on the stove or in the oven. This must be done *very, very slowly,* or your experiment will be in vain. If the salt is heated too much, especially after it dries, it will "pop" out of the container. Lower your heating while there is still liquid present. Finish with the lowest heat necessary to prevent the salt from jumping out.

You can now visually compare the salt in each. There is no doubt about the difference now. Carefully scrape the two pans with a small flat piece of wood. Collect the two salt portions. Label one "control." If possible, have them weighed by your pharmacist. You will probably find that the control has five or six times more salt in it.

This crude experiment can only show you that the ice has less salt in it than the salt water from which it froze. So do not be disappointed if you do not get water which is absolutely fresh.

Try repeating the experiment with sea water. Do you think that you will get better results if you start off with larger volumes of salt water? Will more refreezing remove more salt?

If you have no deep freeze, you may use the ice cube freezer tray in your refrigerator.

Can stalactites
be made at home?

When you go on a long vacation trip by automobile, you will often see a billboard advertising some cavern to visit. Your family is usually not able to resist the urge to explore one of nature's spectacular formations. Everybody talks about the underground trip for many weeks. How would you like to recapture some of this excitement by growing your own honest-to-goodness "stone icicles" called STALACTITES (stuh-LAK-tites)?

Buy a large box of the cheapest form of Epsom salts at your druggist. Dissolve as much of this as you can in a pint or more of hot water. Constant stirring will hasten the process. When cool, pour this strong solution into

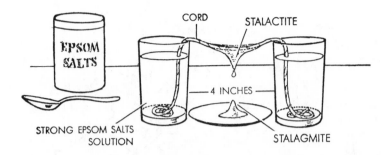

CORD   STALACTITE

EPSOM SALTS

4 INCHES

STRONG EPSOM SALTS
SOLUTION

STALAGMITE

two glasses or jars, almost to the top. Place the glasses about 4 inches apart.

Obtain a cord which will reach from the bottom of one jar to the bottom of the other jar. Do not pull it tight between the jars. Instead, there should be a definite sag in the cord. Place a small colored plate or ash tray directly under the loose cord between the jars.

The cord outside the jars soon becomes wet, and starts to drip at the lowest point. In a few days you will see a building up of one or more stalactites. As these continue to drip, STALAGMITES (stuh-LAG-mites) will be formed in the saucer below. This experiment almost duplicates the process going on in the Carlsbad Caverns in New Mexico, Luray Caverns in Virginia, Howe Caverns in New York, and in many other places.

Of course, the rock found in these caves is not Epsom salts. It is mainly limestone, whose chemical composition is calcium carbonate. This substance is dissolved by acids. To see this at home, you can pour some vinegar (acetic acid) on limestone. You will see bubbles of carbon dioxide as evidence that a chemical reaction is taking place.

The acid found in caves is carbonic acid. It is formed when rain water runs over the ground above the caverns. Rain picks up carbon dioxide from the air, and from decaying vegetation in the soil. When carbon dioxide is dissolved in water it forms carbonic acid. This very weak acid is like club soda or seltzer. It filters through the deposits of limestone. The calcium carbonate rock is then changed by the carbonic acid into a soluble liquid called calcium bicarbonate.

This solution slowly drips through holes in the roof of a cave. Some of the water in the solution evaporates, and some carbon dioxide is also lost. The limestone is no longer able to be in solution, so it comes back as a solid. This deposit builds up the "icicle" from the roof. More solution drips over the stalactite, and the process is repeated for hundreds of thousands of years.

Some of the drippings fall upon the floor of the cave. In this way are formed the "upside-down stalactites" we call stalagmites. Sometimes the upper and lower formations get so big they join to form columns. The colors are caused by various mineral impurities dissolved along with the limestone.

In your experiment you are using a strong Epsom salt solution, instead of dissolved limestone. Epsom salt is magnesium sulfate. The cord gets wet because liquids can rise or move along surfaces or through tiny tubes. It is caused by the attraction molecules have for each other and is called CAPILLARY ACTION.

The sag in the cord allows gravity to pull the drops down to the lowest point. Here some water evaporates, and the Epsom salts do not have enough water to remain dissolved. The stalactites and stalagmites are formed as they come out of solution—just as in caverns.

Try different types and thicknesses of string, rope, or cloth for your bridge. Experiment by tightening the rope between the jars and also with dips of various kinds. What would happen if you tied a knot between the two jars? Will you get colored formations if you dissolve small amounts of vegetable dyes with the Epsom salt? Can you use table salt in this manner to get stalactites?

The next time you walk or ride under a crossroad underpass, look at the bottom of the concrete or stone arch. You may see some stalactites hanging there. They are formed in the same way as in caves.

The concrete and the mortar used to join the rocks contain calcium compounds. These are acted upon by the acid rain water dripping through the cracks. Also look on the sidewalk under some of these stalactites. You may see the start of a stalagmite.

There is a good way to remember the difference between a stalactite and a stalagmite. Just think of the word "tight" in stalactite. That's the way this formation holds on to the roof of a cave.

AIR AND WEATHER

Why do people go to the seashore?

As soon as warm weather arrives, why do folks start thinking about going to the beach? Of course, it is fun to splash in the surf, and to eat all kinds of goodies, and to take thrilling rides. It also gives one a feeling of spaciousness to see the wide expanse of blue water, and the ships sinking into the horizon. But the biggest reason is that the seashore is cool and breezy, while inland the people may be sweltering.

On a warm summer day the ocean is cooler than the land, even though the same amount of sunlight is heating both. In New York City for example, the surface sea water might have a temperature of about 65 degrees Fahrenheit, and would only go up very slightly during the day. Meanwhile the air may be 90 degrees Fahrenheit. The land, too, has a high temperature, since it heats up much faster than the water does.

You can do an experiment to learn whether land or water heats up faster. You must use *equal* amounts of soil and water. Otherwise your comparison is meaningless. So get a large pan of earth and another similar size pan of water. The pans must be made of the same material. Place them in a cool place long enough for both to have the *same* temperature. You may even put both in a refrigerator for a while.

Insert a thermometer into each pan *to the same depth* from the surface. If possible use two previously checked similar thermometers. Place the two pans close to each other in the hot sun. Take simultaneous temperature readings every five minutes of the earth and the water. Readings should be taken in the shaded part of the pans. Otherwise the sun will heat the thermometer in the water directly. Fill in the chart below, until you have enough figures to help you draw your conclusion.

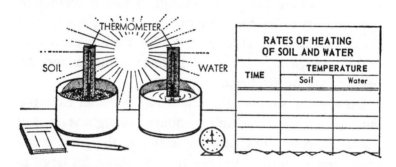

	RATES OF HEATING OF SOIL AND WATER	
TIME	TEMPERATURE	
	Soil	Water

If you wish to do this experiment in the wintertime, place both containers outside the house. When both earth and water have the same temperature, bring them into the warm house. Take their temperatures, both at the same time, every five minutes.

As you showed, it takes longer to heat up water than land. Is the reverse true? Will it take longer for water to lose its heat than an equal amount of earth? This time start with both containers equally warm. Continue as in

the other experiment. Label your new chart RATES OF
COOLING OF EARTH AND WATER.

You will see that water also cools off at a slower rate
than land. That is why coastal cities are warmer in the
winter than inland cities in the same latitude and altitude. The ocean, which still retains its heat from last
summer, acts like a huge radiator for the air near it. In
the winter in New York City, the surface of the ocean is
about 55 degrees Fahrenheit, while the air is much colder.

Coastal cities have a smaller variation of summer-to-
winter temperatures than inland cities at the same altitude and latitude. Look through an almanac listing
monthly temperatures of many cities. Newspapers, too,
often give daily temperatures of many cities. You will
see that in the summer the inland city is usually much
warmer than a city on the coast. In the winter the inland
city is usually colder.

Here are more suggestions for experiments. When
you are in about 2 feet of water, take the temperature of
the surface of the ocean using a thermometer. Then do
the same below the surface. Is there much difference in
temperature? Does white sand lose and gain heat at the
same rate as dark soil? Does water which is stained dark
with black watercolor paint lose and gain heat the way
ordinary water does?

Another reason why it is pleasant to go to the seashore
is that it is usually breezy at the beach. During the day
the wind blows toward the land, and at night the wind
blows toward the ocean. The reason is that the land
heats up and cools off more rapidly than the ocean.

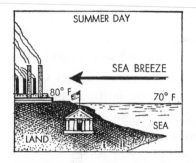

Whenever there is a difference in temperatures between two air masses there is a wind.

In the daytime the land is warmer than the ocean. When the air over the land is heated, it expands and gets lighter in weight. This makes it a zone of low pressure. The cooler, heavier air over the ocean has a greater pressure, and it moves toward and under the rising warm air. This causes a sea breeze.

At night conditions are reversed. The land cools off rapidly when the sun goes down. But the ocean is still about the same temperature as during the day. Now it is *warmer* than the land. There is a horizontal movement of cool air from the land pushing up the warm air over the ocean. This is called a land breeze. Study the illustration.

Most sea breezes start in the late morning and become strongest at about 3 P.M. Do land and sea breezes in your locality have a definite daily schedule? Or do their timetables depend upon the daily announced temperature inland? Are the breezes stronger on a very hot day? Is there a breeze when the temperature inland is the same as the temperature of the ocean surface? How far inland can you detect land and sea breezes? When is there no breeze at all? Is there a change in these breezes during different seasons?

Keep frequent records for many days by observing flags, smoke, or other wind indicators. Make a sufficient number of entries on the following chart, so that you can draw valid conclusions. If there is no wind-speed indicator in the locality, use words such as zero, slight, moderate, fast, and very fast. This type of investigation is best done in the hot summertime.

WINDS AT THE BEACH						
DATE	TIME	DIRECTION	SPEED	TEMPERATURE		WEATHER
				On Land	Ocean Surface	

If you wish, you may include the column for "Temperature of ocean surface." This may often be obtained by phoning the nearby Coast Guard station.

How dusty
is your town?

All over the country, people are waking up to the menace of air pollution. They are becoming vitally concerned as the air over their cities keeps getting more and more filled with harmful dust and gases. These come from many places, but smoke is the biggest source.

Smoke from factories, homes, and automobiles irritate the eyes and throats of many people. Sometimes when smoke and fog mix, a dangerous smog occurs. Then the poisonous dust and gases become so thick that they are often deadly. Victims are mainly people who are physically weak and old people, especially those who have lung and heart trouble.

You can demonstrate how a smog is formed. Get a gallon jug, large soda bottle, or even a milk bottle. Put your lips up tight against the opening and blow hard, building up a pressure in the jug. While blowing, suddenly remove your lips. This sudden lessening of pressure in the jug causes a cooling effect. You may notice that a very small amount of the water vapor in your breath has condensed in the jug.

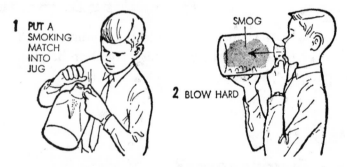

1 PUT A SMOKING MATCH INTO JUG

SMOG

2 BLOW HARD

You are now going to introduce into the jug some very small particles of dust. First light a match and blow it out. While it is still smoking, dip it into the jug so that a very small amount of smoke enters. Again blow into the jug, and suddenly release the pressure as you did before. This time you will see a smog suspended in the jug. The water vapor has condensed into fine droplets of water upon the tiny particles of dust from the smoke.

Experts have estimated that 4 billion particles of smoke are sent into the air by just one puff of cigarette smoke. Can you imagine how loaded the air over a city gets with smoke coming from huge chimneys, and hundreds of thousands of exhaust pipes of automobiles?

However, even clean air always contains some dust. Shine a beam from a flashlight or a film projector across a dark room. Can you see the path of the beam? Shake some clothes or hair into the beam. See how the dust particles reflect the light. Cigarette smoke also makes the beam visible. You can usually see this in a movie theater. Sunbeams too are visible, even in air which is not very dusty.

The type of dust which is *normally* present in the atmosphere is not always considered a menace. In fact, scientists are quick to point out that dust is responsible for the blue skies, colorful dawns and sunsets, and even for raindrops. This is not necessarily the dust which pollutes the air.

The air, as you know, always contains a certain amount of living dust particles, such as pollen from thousands of different plants. Bacteria and mold are also present all the time. Keep a moist piece of bread out in the open, in the dark. Some of the invisible mold dust, called spores, will come out of the air and land on the bread. They will produce different colored molds.

You can collect dust to learn how much settles in different places. A simple way is to place a white piece of paper in various locations free from drafts. Try your bedroom, kitchen, basement, and on a still day, near an open window. For outdoor collections you can lay the paper on the bottom of a deep carton. This may prevent the wind from blowing away the very small bits of dust. You might also use a deep jar with a wide opening.

Can you identify some of the dust by comparing it with dust you make yourself from known sources? How many ways can you make dust? Shake some pollen from a flower, tear material which causes fine lint, etc. Mix them up and try to identify them, using a magnifying glass or microscope. Can you make drawings of many dust particles you can recognize?

Does any dust form an oily smudge when you rub it on paper? Can you dissolve some of the dust in cleaning fluid? Where do you find most dust? Is all dust from all your sources almost similar? If not, can you give a reason? Can you design a dust collector, with some kind of trap which is not affected by winds?

A clever way to collect dust is to use some kind of sticky material to hold the particles. Can you think of some scheme? How about some Scotch tape, with the sticky side facing up? It would be interesting to place many of these around the house and neighborhood and compare results. Another sticky dust-catcher can be made by smearing a light coating of Vaseline, or a similar grease on a piece of cardboard.

Air pollution control bureaus in many cities collect dust in jars of water. In this way they do not worry about wind and rain. The jars are kept on roofs, streets, and near factories for weeks or months. The dust is filtered off and weighed. Of course, this does not indicate the amount of dust that was dissolved in the water. However, evaporation of the water would leave all the dust. Other complicated methods are also used. The report is calculated by giving the weight of dust per square mile for a certain time period.

You can collect dust as the experts do. Use a large wide-mouth collecting jar. Get some large empty mustard jars at your delicatessen store. Put sufficient water in each jar so that it will not evaporate during your collection period. Place some jars on roofs which are near factories, and some on roofs which are in residential neighborhoods. Use homes of your friends. Label each jar. Compare amounts of dust from various sources simply by inspection.

Encyclopedias have much to tell you about air pollution. Librarians have files of clippings on this topic. Find out what a pollen count means to a person having hay fever. Is there an air pollution control bureau in your city? What are some laws in your town which limit smoke production?

Which hair works best in hygrometers?

The invisible moisture in the air is called HUMIDITY. On days when humidity is high we are usually uncomfortable, because our perspiration does not easily evaporate and help cool us. Humidity changes constantly, and people listen carefully to weather broadcasts on TV and radio for the percentage of relative humidity.

They know that when the figure increases it indicates rainy, or at least, unsettled weather. But if the relative humidity goes down, the weather will generally be fair. Of course, knowing only the humidity is not enough to forecast weather. There are many other conditions affecting weather. Nevertheless, it is an important clue.

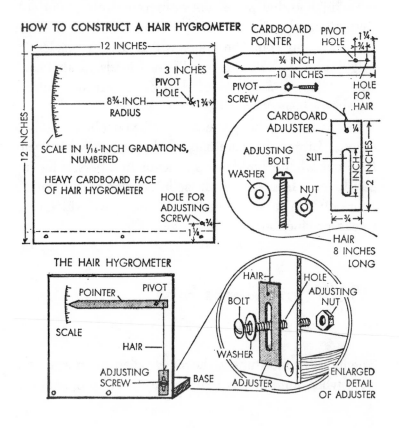

HOW TO CONSTRUCT A HAIR HYGROMETER

The instrument used to measure humidity is the HY-GROMETER (high-GROM-eh-ter). One type uses a hair which stretches slightly when the humidity gets greater. It also shrinks a bit when the air gets dryer. The hair is connected to the end of a loosely pivoted pointer. The changes in hair length make the pointer move up and down on a scale ruled off on cardboard.

Your research problem is to find out which kind of hair is most reliable, and gives consistent, accurate results. Try thick and thin human hairs from blonds, brunettes, and redheads. You will need about 7-inch lengths. Also experiment with horsehair, and even long hair from collies. Each hair must be washed in a detergent, or wiped with a cleaning fluid. This removes the natural oil on it which prevents the moisture from getting to the hair.

The cardboard should be about 12 inches by 12 inches. Ask your druggist for some discarded window displays made of stiff cardboard. These are excellent—almost like plywood, yet they are easily cut by large scissors. The backs are usually white. You might also use sides of strong paper cartons.

The pointer should be about 10 inches long and about ¾ inches wide. The longer it is, the more movement it will make on the scale. It must also have enough weight so that it keeps the hair tight at all times.

The pivot hole is extremely important. *It must be loose.* A good idea is to use a hole puncher for the hole in the pointer. Make another hole in the backboard. Attach the pointer by using a bolt which is slightly smaller than the pivot hole, so that the pointer does not bind at any time. However, the bolt may fit tightly through the backboard. The nut, of course, should not be tightened too much.

Invent other ideas for getting loose pivots. How about putting a bead between the pointer and backboard, and sticking a pin through? Glue the bead in place. Can you use eyelets, washers, or drilled strips of metal? Can you change a paper fastener so that the flat strips do not fray the pointer hole?

The hair should be tied, glued, or Scotch-taped to the pointer and to the slotted adjustment below. See illustration. This simple device saves you work if you make the hair too long or too short. Set the adjustment so the pointer is in the middle of the scale. Then tighten the adjustment nut and do not touch it any more throughout the tests. Instead of the bolt, you may use Scotch tape, staples, or paper clips for clamping the adjustment strip.

The hygrometer must always be vertical. For a stand use a piece of wood which is 3 inches wide, 12 inches long and ¾ inch thick. Attach the cardboard to the edge of the stand with thumb tacks.

The scale should have lines every ⅟₁₆ inch apart. Number them so they can be used for reference. In your notes, have a chart listing the date, time, relative humidity as given on radio or TV, and also the number which

DATE	TIME	ANNOUNCED RELATIVE HUMIDITY	READING ON HAIR HYGROMETER

the pointer indicates. You will find that it will take several days for the hair to get "worked in." After this you will notice definite, correct movements in accordance with the weather reports.

If you wish to get true outside readings with this hygrometer, you must keep it indoors near an open window in the warm weather. In the fall and winter, place it outside on a window sill. Devise some way of protecting this delicate instrument against dust, jarring, wind and rain.

If treated carefully, this homemade hygrometer can be fairly accurate. It may not give you the same relative humidity as the weather report, or another commercial hygrometer you have at home. But it should go up and down with the announced relative humidity.

If you can make a few hygrometers you could speed up your research. You might also experiment with vegetable strings, such as cotton and linen. Strangely enough these shrink when they get moist, and elongate when they dry—the opposite of animal hairs! See what happens to silk thread (which comes from an insect). Also try gut strings used in some musical instruments.

Do the hairs stretch after several months, so that you cannot get similar readings on your scale for the same announced relative humidity? Does increasing the length of a hair do anything for the sensitivity of your hygrometer? Does the weight of the pointer influence your hygrometer's sensitivity? Try tying small weights to the end of the pointer.

What force can a tornado exert?

Tornadoes are the most destructive of all storms. It is estimated that the horizontal winds inside those funnel-shaped clouds can often reach speeds of 500 miles per hour. That is why you can see very strange sights after tornadoes have passed. Automobiles, buildings, and uprooted trees are often lifted and transported to new locations hundreds of feet away.

Chickens have had all their feathers completely plucked out. But the most convincing evidence of the wind's force is a piece of straw imbedded in the trunk of a tree. You can duplicate this feat at home to the amazement of your onlookers by driving a paper drinking straw into an unbaked and unpeeled potato.

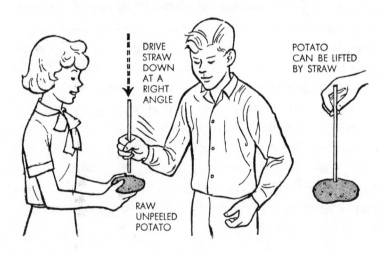

DRIVE STRAW DOWN AT A RIGHT ANGLE

RAW UNPEELED POTATO

POTATO CAN BE LIFTED BY STRAW

Grasp the straw firmly in your fist, but do not crush it. Hold it about 10 inches above the potato and make a fast stab at it. Be sure that the straw strikes the potato at right angles. You will be pleasantly surprised to see the soft straw enter the firm potato. For dramatic effect, you can pick up the potato with the straw.

Penetration was due to INERTIA (in-ER-shuh), which is the tendency of objects to continue what they are doing. The potato had inertia of rest. It remained in the same position for the brief instant that the weak, but fast-moving straw struck it. The moving straw possessed inertia of motion. It continued to move in the same direction it was going, even though it struck the potato.

Experiment by holding the straw in different positions. Try placing your forefinger over the top end of the straw. Does the air trapped in the straw make the straw more rigid? Or does it make little difference in the piercing ability? Try to penetrate different fruits and vegetables, such as unpeeled apples, onions, and radishes.

Do you think that the fibers in celery will not allow this penetration? Or will inertia take care of this obstacle also? Can you get the straw into the hard coverings of cantaloupes or other melons? Can your weak straw pierce balsa wood—if you make the correct fast motion? Will all the successful demonstrations work equally well if, instead of a drinking straw, you used a straw from a whisk broom?

SOUND AND MUSIC

Why do some things
"talk back" to you?

Your family and friends will think you are behaving oddly when you do some of the following experiments. But you can easily prove your sanity with your scientific explanations and soon everybody will join in the fun.

You need a piano—yours or your neighbor's. If possible, open the lid which exposes all the strings. Press the loud pedal. Sing a loud, even note close to the piano strings for several seconds. When you stop, listen carefully to the piano as it sings back to you.

Find out which string has been made to vibrate by your note. You can do this by ear or by touching the suspected string gently with the fingers. You will probably find several strings humming back at you. Sing many separate notes up and down the scale. In each case find the vibrating strings. Sing soft and loud. How do the strings answer?

Shout into the piano. Do more piano strings vibrate? Strike a dinner bell, a thin drinking glass, or anything else which produces a musical sound. What does the piano answer when several people sing loud harmony notes together and suddenly stop? By the way, these experiments will work even if you cannot open the lid to get to the inside of the piano. The strings will answer from within.

The piano has many lengths and thicknesses of wire. These have been tightened so that they will vibrate a definite number of times per second when struck. But a string will also vibrate without being struck. Anything vibrating near it *at the same rate* as its own natural vibration rate, will set a string into motion. This is called SYMPATHETIC VIBRATION. It is what is happening to the piano strings in your experiments.

How does this happen? How can your own vibrations start vibrating the tight strings? Have you ever tried pushing a heavy person on a swing? It is not easy at first, but you time yourself, so that your pushes are made when the person is at the height of his swing and starting to move away from you. Each time, at just the right moment, you give the swing a tap. Soon the swing is high up in the air.

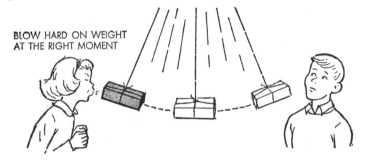

BLOW HARD ON WEIGHT
AT THE RIGHT MOMENT

You can also illustrate this by suspending from a string a very heavy book or other object. Blow on the weight in gusts to get it started. Time your blows so that they occur only when the weight is closest to you and starting to move away. After a while, the heavy weight is swinging high.

Actually, you are now vibrating in tune with the vibrating swing or weight. You have demonstrated what a scientist calls RESONANCE (REZ-uh-nence). It is a re-enforcement of sound caused by sympathetic vibration.

Every object has a natural frequency or period of vibration. A diver bounces up and down on a springboard until he is in "tune" with its natural frequency. In this way he gets a greater bounce than if he were "out of step." Sometimes, while the kitchen refrigerator is working, a dish inside will rattle. This often happens because the electric motor may be producing vibrations which are in resonance with the natural frequency of the dish.

Certain musical notes from your radio or record player often resonate with the wooden cabinet, and cause a rumble. Sometimes a vase or some trinket on a shelf will bounce around. It has been said that a powerful operatic tenor's voice can shatter a drinking glass by sympathetic vibrations. He sings and holds many true notes until he reaches the natural period of vibration of the glass, and it shatters.

Another experiment with resonance is done with frying pans, baking pans, broilers, and other metal containers. Hang each from a string, and sing a steady loud note into the pans for a few seconds. When you stop, you will hear the metal repeat your note because of the sympathetic vibrations. Try many types of pans. Very heavy iron pans resonate excellently for certain notes.

Put a sea shell to your ear. The so-called sound of the waves is caused by the slight noises in the air, which are in tune with the shell. Sympathetic vibrations cause the air in the shell to resonate (make the sound louder).

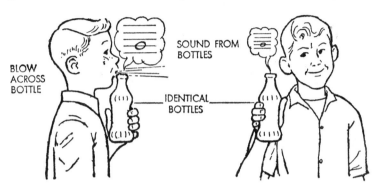

Have a friend hold a soda-pop or milk bottle at his ear, keeping the bottle opening unobstructed. Blow across the opening of another *similar* bottle until you cause a sound in your bottle. Your friend should now hear the same tone produced in his bottle too. The vibrations set up in one bottle caused sympathetic vibrations in the other identical bottle.

Another experiment is to go close to a piano, holding a milk bottle to your ear. Strike different notes up and down the scale. You will soon discover one piano note which causes the air in the bottle to "talk back" loudest. The air in the bottle is now vibrating at the same rate as the piano string, and resonance occurs.

Can you produce
animal sounds by vibration?

With a few simple household materials you can assemble a noise producer which can imitate a rooster's crow, a dog's bark, or even a lion's growl. You can also use this device for answering many questions about the science of sound.

Make your first model from the bottom 4 inches of a quart-size paper milk container. Punch a small hole in the center of the bottom and thread the end of a strong 24-inch string through it. Tie many knots in the end of the string which is in the container, so that it cannot be

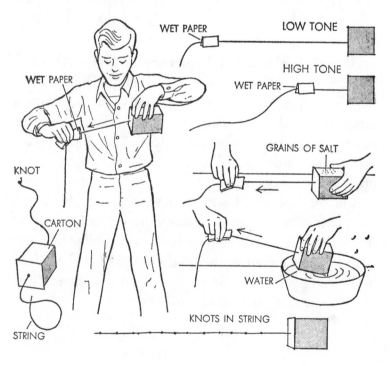

pulled through. What you now have is really half of a "string telephone."

Wet a piece of paper toweling, napkin, or cloth. Squeeze out the excess water so that the material is moist, but not dripping. Now hold the noise maker in your left hand. Place the moistened paper or cloth between your thumb and fingers of your right hand. Grasp the string between the wet material and while pressing it with the fingers, give a quick pull. The moist paper will alternately skid and grip and "chatter" along the string.

This vibration in the string will produce a squawking noise. The sound is made louder, of course, because the bottom of the container vibrates and the box amplifies the noise.

Practice varying the pressure and the speed of the moist paper along the string and you will soon be able to imitate recognizable animal sounds. An even pressure and pull will make an almost musical, pleasant sound. A jerky motion, with unequal pressures, will cause an annoying noise.

You can feel the vibration in your fingers. If some salt or sand is carefully placed on the side of the milk carton while it is on a table, you can see the grains dancing about as you vibrate the string. You can also make water ripple and splash, because of the vibration. Simply dip a small part of a corner of the box into a basin of water, while vibrating the string.

If the fingers pull the string close to the box a high tone is produced, since only a small length of the string is vibrating. If the pull is started near the end of the string, a low tone is produced because a longer string is now vibrating.

Place the box against a rug or coat and the sound is deadened. Test many materials to discover which are good sound insulators. When the box is placed against the door of a room or a closet, or on a plate-glass window, the vibration is greatly amplified as in a violin.

Can you cause the string to vibrate if it is dry? Try pulling a dry string with the fingernail of the thumb rubbing against the string. Another scheme is to rub rosin (used for violin bows) on the dry string and then pull it with the fingers. Still another dry method is to make knots in the string and then pull the fingers along the bumps. Tie knots close together, evenly spaced. In another string, tie knots farther apart. Which gives the higher tone?

Experiment with different sizes of boxes; oatmeal boxes, coffee cans, large pails, cardboard boxes, ½-gallon ice-cream containers, and also larger ones.

You will get a very pleasant sound with larger containers by plucking, instead of rubbing. You will have to hold the container on the floor between your feet and hold the string tight.

Try different kinds of string. You will find some are much better than others. Different effects are produced too by changing the way the end of the string is attached to the box. Instead of knots, try threading the string through a large button, or tie it around a match stick or a metal washer.

Please have mercy on the family!

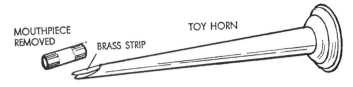

MOUTHPIECE REMOVED

BRASS STRIP

TOY HORN

Are you a
sodastraw musician?

A bunch of soda straws and sharp scissors are all you need to put you into the music business. Needless to say, you will not be able to compete with this country's big-name orchestras. But you can certainly wile away many hours of exciting experimentation.

Have you ever removed the wood or plastic mouthpiece from a toy New Year's party horn? Under it there is a loose strip of brass, which moves rapidly back and forth when you blow into the horn. The movement also causes the column (tube) of air in the horn to vibrate. This produces the familiar horn sound. The tone depends upon the size of the strip of brass and also upon the length of the air column. This is the principle of the soda-straw instrument you are going to make.

First, pinch flat about ¾ inch of one end of a soda straw. Cut off the corners as shown in the illustration. Now put this end of the straw far enough into your mouth so that your lips do not touch the cut corners. Close your lips around the straw and blow hard. You should get a musical sound.

If no sound is produced, the flattened opening of the end of the straw in your mouth is either closed or open

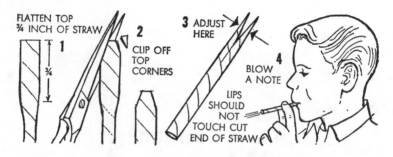

too much. Gently pressing the sides with the fingers will widen the opening. Pressing the flattened sections will close the opening. Three or four adjustments will give you the idea. Sometimes all you may get is a squeak. This is often caused by a rough edge left by a poor cut, especially with dull scissors. In such a case, simply cut two new corners.

The long straw makes a low-pitched tone. While you are blowing, use the scissors to cut off half-inch pieces from the end. As you shorten the straw you will hear the musical tones going up the scale. See if you can play all the notes of a musical scale by cutting the straw at pencil marks. The placement of these marks was obtained from trials on other straws. (Hint: In your experimentation, always cut off *less*, rather than more, until you tune in on the correct note.)

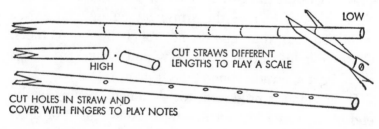

Hold a series of tuned straws of different lengths in your hands. Blow the proper tones to play a recognizable melody. To form an orchestra, get a group of friends in front of a "leader" who knows music. Each person should hold tuned straws and blow only when he is told. He then holds the notes for as long as the leader indicates.

Experiment with many kinds of straws and also of all diameters and lengths. Try plastic straws and "heavy-

duty" straws used in hospitals and some ice-cream parlors. When mouthpieces of straws get too moist, do they lose their stiffness and tone? Will it do any good to waterproof them with clear nail lacquer or shellac?

What is the effect of placing funnels of various sizes at the end of the straw? Does the tone change in pitch? Does the sound get louder?

Slip a long, unprepared, wider straw over the one you are using. By moving it forward or backward like a trombone, will you get different tones? Another experiment is to cut small round holes an inch apart in a straw. Can you vary the tone while you are blowing, by placing your fingers over different holes?

Will you get the same tones from similar kinds of straws if their mouthpieces are cut the same way? If the corners are cut longer or shorter? Do all straws of the same length give the same tone even if one is wider than the other? Can you blow two differently tuned straws in harmony if placed in the mouth at the same time?

Examine professional instruments. See how they change the length of their vibrating air columns. For example, in a trombone the slide moves in and out, while in a saxophone valves are opened and closed.

In another experiment you can again use a straw to show that there is a change of pitch when there is a change in length of air tube. With a razor blade, carefully cut halfway across a soda straw, about 2 inches from one end. Bend the straw at right angles as shown in the illustration.

Put the longer end in a glass of water. Gently blow through the short end and you will produce a soft whistle. You may have to press together the top of the tube, which is in the water. If you raise and lower the water while you are blowing, you will get an interesting change of tones. Try to play a simple tune.

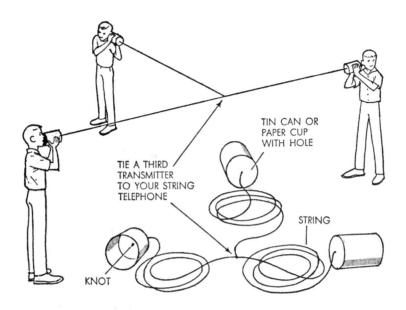

TIN CAN OR
PAPER CUP
WITH HOLE

TIE A THIRD
TRANSMITTER
TO YOUR STRING
TELEPHONE

STRING

KNOT

Will a string telephone "party line" work?

Construct a regular string telephone having a paper-cup or tin-can transmitter at each end, so that two people can use it. To the middle of this string, tie the end of another string telephone having only one transmitter at the other end. Stretch it out at right angles.

Can a third person talk and listen in at this extension telephone? How many additions are possible? Does each addition weaken the vibrations? How far can you hear?

CHEMISTRY AND ANALYSIS

Where is carbon dioxide found in your home?

Carbon dioxide is constantly being given off into the air by the breathing of people and other animals, by plants, burning of fuels, and decay of all kinds. In addition, many household activities use this gas for different purposes. How can you act as a scientific detective to find this colorless, odorless, and tasteless gas?

Chemists have made this task very simple for you. They use a very inexpensive solution called LIME WATER, which looks like water. But shake it up with some carbon dioxide gas, and it turns milky because of the formation of a very fine chalklike powder.

You can make lime water by adding a spoonful of SLAKED LIME to a quart of water. Shake, and then allow the excess lime to settle to the bottom. This may take several hours. Carefully, pour off some of the clear lime water from the top into a labeled bottle with a *tight* cover or cork. Use it sparingly.

You can get the tiny amount of lime you need from any of your friends who own homes, since it is widely used on lawns. Lime is also used in new buildings where plastering is done. It mainly comes in large bags. Since it is so cheap, perhaps a friendly gardener, masonry-supply or hardware dealer will give you a tablespoonful. Hobby shops may sell it for chemistry-set refills. Any older friend studying chemistry can help obtain it. But if you still have difficulty, ask any druggist for a 20-cent bottle of lime water.

Here is a short lesson in chemistry to help you understand how the lime water test works. Carbon dioxide is formed when carbon chemically combines with oxygen. It is called CO_2 because one atom of carbon combines with two atoms of oxygen. Lime water is a solution of calcium hydroxide. When it reacts with CO_2 it forms calcium carbonate. This white powder cannot be dissolved in the water and so it settles to the bottom. A chemist would say a PRECIPITATE (pre-sip-ih-tate) is formed.

You can discover the presence of CO_2 in dozens of places at home. First try the suggested experiments. These will help you search for other sources of the gas.

Pour a small amount of clear lime water into a test tube, plastic pill bottle, or narrow bottle. By means of a drinking straw, blow into the solution. You will see the milky color develop, proving that you are alive. Some of the food you eat contains carbon which eventually becomes part of your body. The air you breathe contains oxygen, which is carried by your blood to the smallest part of you called a cell.

In the cell the carbon combines with the oxygen in a rather complicated manner, forming CO_2, a process called OXIDATION (ox-ih-DAY-shon). The CO_2 gas gets

81

BLOW INTO A
TEST TUBE OF
LIMEWATER

LOWER
BURNING
CANDLE
INTO JAR
OF LIME-
WATER

Limewater

into your blood and finally into your lungs, and you blow it into the lime water.

If you blow too long the precipitate dissolves and the lime water clears up again. This is perfectly normal. Excess CO_2 makes the solution acid and the calcium carbonate is dissolved. It forms calcium bicarbonate, which is soluble.

Continue your experimentation by pouring a little lime water into a jar. Wrap a wire around a candle so that when you light it, it can be lowered into the jar without burning your fingers. When the flame goes out (or even before it does) swish around the lime water.

A candle is made of paraffin which contains carbon. During burning, which is a fast oxidation, CO_2 is formed. Oxidation produces heat. Our bodies are kept warm by slow oxidation.

Wood, paper, and cotton are made of CELLULOSE (SELL-you-lowss), which contains carbon. Test wooden and paper matches and also some cotton by burning them as you did the candle. Also make suitable holders and burn bits of bread, butter, and many other foods to test for production of CO_2. Do these tests convince you that most foods contain carbon in their chemical composition?

Is there CO_2 in air? Pour some lime water in a glass and allow it to stand overnight. If the test solution turns milky, where is the CO_2 coming from? Do you see why your bottle of lime water must be tightly covered? Chemists often put Vaseline around the stopper so that the lime water will stay fresh longer.

Put a tablespoon of baking *soda* (bicarbonate of soda) into a jar. Pour in some vinegar. The reaction which you will witness is producing CO_2. Devise some way of testing for the gas. Can you lower a pill bottle containing lime water into the jar?

When baking *powder* has only water added to it, a reaction occurs which produces carbon dioxide. Test for CO_2. This gas is used in baking in order to "raise the dough." The CO_2 bubbles will mix with the batter and expand during baking. This makes the holes in bread and the fluffiness in cakes.

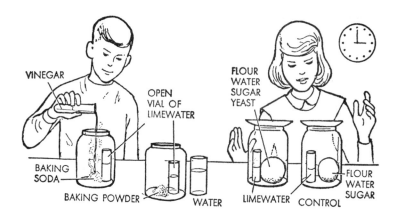

Yeast is also used in baking, for the same reason. Test for CO_2 as follows: Mix some yeast, flour, and a pinch of sugar. Put this dough into a jar. Cover the top with a saucer, and keep in a warm place for several hours. The yeast grows rapidly and acts on the sugar. This produces large quantities of CO_2 gas among other products. In the jar you might also place an uncovered pill vial containing lime water.

At the same time set up a similar jar containing an open vial of lime water, but without the yeast. See if CO_2 is now produced. This is your control, which you will use for comparison. This is necessary because if your control does not show the presence of CO_2, you are now certain that it is because the yeast was not added.

Test for CO_2 when Alka-Seltzer or Bromo-Seltzer and similar bubble-producing tablets are placed in water. Also test soda water for CO_2. Open a *very cold* bottle of soda pop. Immediately place a balloon over the opening. Allow the bottle to get warm, shaking and striking it from time to time so more CO_2 is freed.

Remove the balloon, holding its opening closed, and in

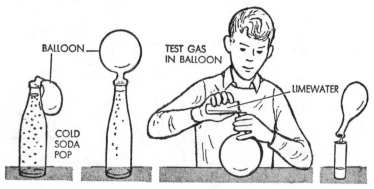

BALLOON
COLD SODA POP
TEST GAS IN BALLOON
LIMEWATER

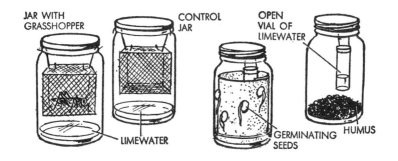

JAR WITH GRASSHOPPER
CONTROL JAR
OPEN VIAL OF LIMEWATER
LIMEWATER
GERMINATING SEEDS
HUMUS

an upright position. Pour a little lime water into the balloon. Swish the test liquid around and pour it back into a small glass. Is it milky? See if you get the same results if you put a balloon over a bottle of tap water.

Does a grasshopper give off CO_2? Put one into a tiny cage suspended in a jar. Pour some lime water on the bottom and cover the jar. As a control, have an identical setup—but without the grasshopper.

Use the same kind of arrangement to find out if seeds give off carbon dioxide when they first start to grow (germinate). You might also find out whether humus gives off CO_2. Humus is the decayed animal and vegetable matter found on the ground in forests.

Can you prove that dry ice is frozen CO_2? Get a piece from your ice-cream man. Do not handle with bare hands!

Lime water is a useful scientific device for opening many doors to interesting experiments. Devise some CO_2 research of your own. You can read a great deal about this important gas in every encyclopedia and in many science books.

85

Can plants be used to make good dyes for cloth?

Historians tell us that people in almost all lands have been dyeing fabrics for many thousands of years. They find proof of this in the Egyptian tombs and in ancient ruins of China, Persia, and India. Most dyes were made by crushing and soaking certain parts of plants, such as leaves, flowers, stems, roots, bark, shells of nuts, berries, and skins of fruits and vegetables.

Today, the chemist knows the formula of the active chemical found in each dye. Since the middle of the nineteenth century the industry has been busy obtaining new dyes from coal tar and also inventing other synthetic coloring materials. Hardly any dyes are made from plants today.

But you can make your own dyes. Then you will recapture some of the excitement felt by the early settlers in America when they proudly dyed their crude fabrics. You will have fun selecting your coloring materials from a large assortment of plants found in your home garden, vegetable store, and in nearby fields and woods.

Here are some suggestions for possible sources of dyes: beets, carrots, onion skins, rhubarb, spinach, tea, coffee, tobacco, sumach, bark of red oak, willow and other trees. Experiment with colorful flowers such as black-eyed Susan, goldenrod, marigold, dahlia, petunia. You might also try blueberries, raspberries, elderberries, and the outer coverings of butternut, hickory, and other nuts.

In general, the dye is made by chopping up into small

MAKING THE DYE
CHOP OR GRIND VEGETABLES

BOIL SLOWLY 1 HOUR

USE GLASS OR ENAMEL PANS

STRAIN OUT VEGETABLES

pieces, or thoroughly grinding, the part you intend to use. Place in a small quantity of water and boil slowly for about an hour. Add water to replace any which boils away. Strain the dye through a cloth in order to remove plant materials. Dip the cloth to be dyed into this dye bath and boil for about fifteen minutes. Stir constantly with a stick. Remove the cloth, rinse it, and dry it in the shade.

You can change any of these conditions and note the effect on the dyeing. Try adding more water so the dye bath is weaker. Does boiling for a longer period make a more deeply colored dye? Will the cloth be dyed if it is dipped into a cool dye bath? Does filtering the dye through a paper towel give more even results?

Each dye should be tried on different kinds of white cloth. It is very important that first all starch and other stiffeners be removed. Wash in hot water and detergent and rinse well. Cut 3-inch squares of cotton, wool, linen, silk, rayon, nylon, and even Fiberglas. Label each one carefully. If possible, use a ball-point pen containing laundry marking ink. Save the samples for future reference. Keep records in a notebook and draw your own conclusions from your observed results.

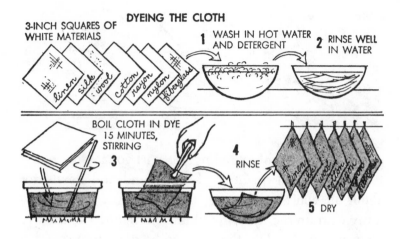

DYEING THE CLOTH

3-INCH SQUARES OF WHITE MATERIALS

linen silk wool cotton rayon nylon fiberglass

1 WASH IN HOT WATER AND DETERGENT

2 RINSE WELL IN WATER

BOIL CLOTH IN DYE 15 MINUTES, STIRRING

3

4 RINSE

5 DRY

Find out how permanent or "fast" each dye is by rinsing part of each sample, first in cold and then in hot water. Note whether the color runs. Keep one unrinsed part of the cloth as your control, for comparison purposes. Another test might be to see whether sunlight bleaches the dye.

You will probably find that fibers such as silk and wool, which come from animals, will take dyes well. They are protein and react chemically with the dye to form colors which are not easily washed out. Cloth made from vegetable fibers can be dyed in a more permanent manner by first being treated with a chemical called a MORDANT (MORE-dent). However, in practice it is best to use a mordant for all fabrics.

One such mordant is alum. You may have some at home. If not, buy a small amount at the drugstore. It is very inexpensive. When a solution of alum has some household ammonia water added to it, a jellylike chemical called a GEL (JELL) is produced. This is aluminum hydroxide and it clings to the fibers of the cloth. It also attracts the dye, holding it so that it does not wash away.

Make up the alum bath by stirring about half a teaspoonful of alum into half a quart of water. Label it. In a separate container, prepare a half quart of diluted household ammonia. Label this too.

When ready to use the mordant, dip the white cloth into the alum bath for a few minutes. The alum penetrates into the fibers. The cloth is then dipped into the diluted ammonia water for a minute or so. This forms the aluminum hydroxide gel which holds dye so well. The cloth is now ready to be dipped into the dye which is used *cold*.

You may try varying this procedure. Make one mordanting bath by adding the alum to the ammonia bath. Then you simply dip the cloth into this combined bath before dyeing.

The ammonia odor may annoy you, so work near an open window where the draft wafts the ammonia *away* from you. However, instead of ammonia water, you may use a weak solution of washing soda.

It is best not to use metal pans. Use glass, enamel, or stainless steel.

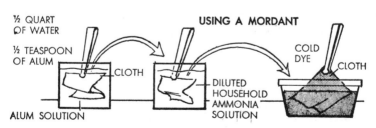

½ QUART OF WATER
½ TEASPOON OF ALUM
CLOTH
ALUM SOLUTION

USING A MORDANT
DILUTED HOUSEHOLD AMMONIA SOLUTION

COLD DYE
CLOTH

Make charts containing your dyed samples and list what you used. This is an excellent science-fair exhibit.

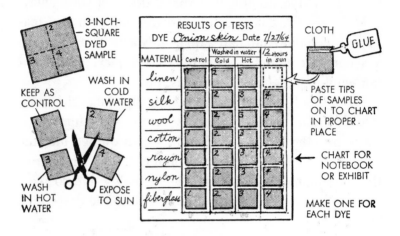

How much of the air is oxygen?

You have heard that fuels need oxygen in order to burn. Yet only a part of the air is oxygen. If the air were almost all oxygen, every time you struck a match you would be holding a torch! We should be thankful that most of the air is nitrogen, a colorless gas that quenches fires.

You can find out the proportion of oxygen in the air by means of an experiment. You will need a pan and two of the tallest drinking glasses you can find. They should have straight sides, not sloping ones. Large olive jars would be much better.

Obtain a small clump of steel wool which is not combined with soap, and has not been treated with a rust preventive. Wash the steel wool with a detergent in hot water. Rinse well but do not dry. Spread the steel wires apart so that they take up more space. Wedge a small bundle into the bottom of one glass so that it will remain there when the glass is turned upside down.

Turn over the empty glass also and place both inverted glasses in a pan containing about 2 inches of water. Notice that the water level is at the bottom of each glass. But in a short time, the water will start entering only the glass with the steel wool. Allow to remain in this position until the water stops rising. This may take a day or more.

During this time, because of the moist condition inside the glass, some of the steel is turning to rust. Rust is iron oxide, and it is formed when iron or steel combines with oxygen. The oxygen is taken from the trapped air in the inverted glass. As the oxygen is used up, the air pressure outside the glass forces the water into the glass.

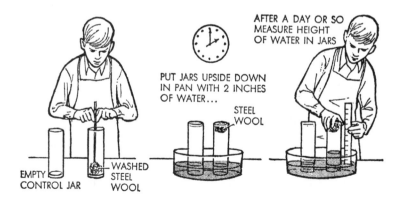

The height of the water in the glass indicates the amount of oxygen that was removed by the rusting steel.

Carefully, lift the glass to just *below* the water level in the pan. Do not let any water escape from the glass. Measure the height of the water in the glass. Compare it with the height of the glass. It should be about one-fifth, since one-fifth of the air is oxygen. Actually, 21 per cent of the air is oxygen.

Examine your control. Did the water rise in the empty glass? Repeat the experiment several times and average your results. Record your results as fractions. Or if you multiply your fraction by 100 you will get your answer in per cent.

$$\frac{\text{Height of water in jar}}{\text{Height of jar}} \times 100 = \% \text{ oxygen used up}$$

Test whether the gases remaining in the glass allow something to burn in it. Carefully, place a small saucer under the inverted glass while it is still in the water. Press the saucer against the glass and invert it quickly with the water in it. Do not allow outside air to get in. Light a match and insert it into glass. Why does the match go out immediately?

Here are some research questions to answer when you repeat the experiment. Will the water rise higher if you use more steel wool? Keep a record of the speed of the rising. Does the water rise evenly or faster at the beginning? Will heat increase the rusting speed and lessen the time to complete the experiment? You can heat the glass by shining an electric light close to it. Or you may put the pan in a warm place.

Are you able to copy newspaper print?

When you were small you may have spent many happy hours transferring newspaper print to a blank sheet of paper. First you rubbed a candle evenly over the newspaper picture or type. Then you laid it flat, placed a blank piece of writing paper over the waxed section, and rubbed it smoothly. If you did a good job you were rewarded with a reversed copy of the picture. The reason was that the wax had picked up some of the printer's ink and transferred it to your copy. But usually the results were rather poor.

By doing some careful experimentation, can you now improve these results? Can you make better copies of newsprint by developing a copying liquid and perfecting your technique?

Printer's ink is really a form of paint. Painters use turpentine to dissolve paint. Suppose some turpentine were brushed on a newspaper. Would some of the print come off on a piece of blank paper which is rubbed over it? Try it. Nearly every home owner has turpentine. Artists who do oil paintings also have some on hand. You only need a very small amount.

You will discover that to get good impressions you need a very smooth surface upon which to rub the papers.

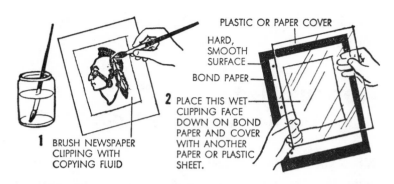

1 BRUSH NEWSPAPER CLIPPING WITH COPYING FLUID

PLASTIC OR PAPER COVER
HARD, SMOOTH SURFACE
BOND PAPER

2 PLACE THIS WET CLIPPING FACE DOWN ON BOND PAPER AND COVER WITH ANOTHER PAPER OR PLASTIC SHEET.

You might use a cooky sheet, an old porcelain table top, smooth wood (not your furniture), or a piece of glass from an old picture frame.

There are several ways to set up the papers for rubbing. Use a rag or a brush to spread your solution on the newspaper picture you wish to copy. Place this face down on a blank piece of writing paper which is already in place on your rubbing surface. Now put another piece of paper on top of the newspaper and rub evenly with your fingers, palm, roller, or spoon.

Another method is to put the solution on the newspaper. Then put it face up on the smooth surface. Place your blank paper over it. Put another piece of paper over this and use it for rubbing. You might decide to use a sheet of light plastic or cardboard over everything and upon which you will rub. This will save many sheets of paper. Use small pieces of paper for your many tests.

You will probably discover that the idea of using turpentine alone sounds good, but it does not work well. The following formula works much better: Pour half a pill bottle of turpentine into a small jar. Add four times this amount of water. The turpentine will float to the top

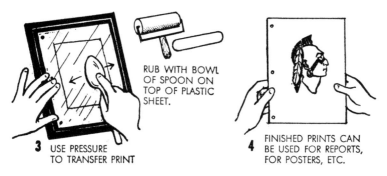

RUB WITH BOWL OF SPOON ON TOP OF PLASTIC SHEET.

3 USE PRESSURE TO TRANSFER PRINT

4 FINISHED PRINTS CAN BE USED FOR REPORTS, FOR POSTERS, ETC.

because it does not mix with water. But if you will add a small pinch of laundry detergent or soap and mix thoroughly, the turpentine will not separate any more.

Experiment with this copying fluid. Add more water or make up a new formula with less water. Keep accurate records of your formula and the results. Do not depend upon your memory. Above all, only change *one* condition at a time for each trial. For example, if you add one more part of water to your formula and you wish to compare results, you must have everything the same as for your last trial. You must do the same amount of rubbing, use the same papers, make the print as wet as before, etc.

You will finally arrive at the best formula for the technique you are using. Of course, your pictures will come out reversed. They will read correctly only in a mirror. Can you think of a way of making unreversed copies?

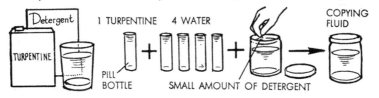

Detergent TURPENTINE 1 TURPENTINE 4 WATER COPYING FLUID PILL BOTTLE SMALL AMOUNT OF DETERGENT

Can color be copied as well as black? Do all newspapers give the same results? Does it make a difference how long you press or how wet you make the newspaper? How many copies can you get from one master?

This kind of research illustrates how a scientist often has to work to get results. It will take lots of time, but you will enjoy every minute of it. Keep your work area neat by covering it with newspapers. Discard your rags and wet paper outside, to eliminate odors.

When does a negative look like a print?

By means of chemistry you can reverse the appearance of a black photo negative so that it looks almost like a print. A negative consists of very tiny silver particles, embedded in the gelatin which is coated on the film. The silver in this condition looks black.

An iodine solution which you have in your medicine chest can change the silver into silver iodide. This gives the negative a light brown color. When cleared and viewed against a dark background the negative will almost resemble a black and white print.

Place the negative in a small tray or soup plate, so it lies flat with the dull side up. Pour over it enough tincture of iodine to cover it. Swish the solution around so that the negative is always covered. Do this for about fifteen minutes. You will see the black change to brown.

Remove the negative by grasping an edge with tweezers or tongs. Wash it in water and then keep it in some

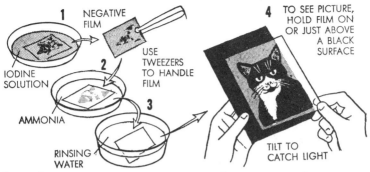

1 NEGATIVE FILM

IODINE SOLUTION

2 USE TWEEZERS TO HANDLE FILM

AMMONIA

3

RINSING WATER

4 TO SEE PICTURE, HOLD FILM ON OR JUST ABOVE A BLACK SURFACE

TILT TO CATCH LIGHT

household ammonia water until the excess iodine is removed. Now rinse away the ammonia. You can view the film immediately by holding its edges with the tweezers. Or you can place the negative on a cloth or blotter until it dries.

To see a positive image clearly you must have a dark background some place behind the film, either close or farther away. Tilt the film in all directions. You will discover the best angle for the light to shine on the negative so it will look like a print. Try looking at both sides.

Best results are obtained from sharp, well-exposed negatives, having large faces or figures in it. Experiment by varying the time the negative is acted upon by the iodine and the ammonia.

Instead of the tincture of iodine you can ask your druggist to make up a weak solution of water, iodine, and potassium iodide. It is called Lugol's solution and is used for testing food samples for starch. Your science teacher probably has it.

Needless to say, do not contaminate the solution of iodine antiseptic needed by the family for emergencies. Use your own solution. Can you reuse your solution for processing other negatives?

PLANT INVESTIGATIONS

Can you prepare a timetable for nature?

If you are a walker who enjoys observing nature you will get special pleasure in the springtime. On all sides trees and flowers are awakening according to a definite schedule. Each kind of tree or flower has a different date for unfurling its buds. You can make a calendar showing the dates of appearance of each species. This activity is guaranteed to teach you a great many things about these plants.

You will learn to be especially observant, for now you will be looking for something specific. This is good, for every scientist must be observant. During your searching you will discover many other things to interest you. This means you will see more. And as you see more, you will find more things to interest you. In this way your ever-widening circle of knowledge grows larger.

You will learn the names, since you cannot call a tree or flower by number. You will begin to notice where each plant seems to prefer to thrive. You will start reasoning about plants and developing skill in organizing ideas in your mind. Above all, you will remember most of your observations and facts, because you learned them yourself from the best teacher in the world—Nature!

There is a regular procession of leaf unfurlings on trees, and blooming of flowers. For your first records, include only the most obvious and most numerous trees in your locality. Do not attempt to make a thorough census or survey. This is a big, confusing job for a beginner. But once you start on your timetable project you will immediately become aware that on certain trees the buds "pop" days and weeks before other neighborhood trees. Your task is simply to record the order of this appearance.

The names are quite easy to learn, for many home

owners can tell you the name of the large trees on their property. Take a walk with a person who knows trees and you will quickly learn those you need for your calendar. Boy and girl scout organizations know neighborhood trees. So do members of garden clubs, college biology students, local florists and gardeners, Park Department men, and teachers.

There are many excellent books on trees in the library. These have clear pictures showing the shapes, bark, leaves, fruits, and other simple guides to identification. No doubt there will be some trees which you cannot name. Forget about these. Even experts have trouble identifying everything.

Buds on trees are marvels of compactness. They contain folded leaves and flowers and are formed during the growing season of the last summer. In this way, they have the advantage of getting an early start in the spring. The time of opening depends upon a necessary resting period in the cold, and the proper warm temperature in the spring.

As a rule, most trees of the same kind open their buds within a few days of each other. Here and there you may find exceptions. Can you decide the reason for this; size of tree, nearness to a stream, deep shade in the woods, open sunlight in the field, damaged trunk? But if you record a sufficient number of similar trees, a pattern will become obvious.

List your observations in your notebook. You may keep a separate page for each kind of tree, or you may copy all trees as you find them.

POPLAR

DOGWOOD

BEECH

PUSSY WILLOW

TREE	LOCATION	Date of Bud Opening	REMARKS

Later you can study your notes. Decide upon the best average date for the opening of the buds for most trees of a certain kind. Make your calendar on graph paper.

City____ State____	BUD OPENING DATES						Latitude____ Elevation____				
TREE	APRIL						MAY				
	5	10	15	20	25	30	5	10	15	20	25	30

Do the same trees obey the same opening schedule every year? Does a cold or warm winter make any difference? How late or early was the spring weather? Did it affect the buds? Do similar trees in another neighborhood far away obey your tree calendar? Ask a distant friend to help you by preparing one also. Compare notes.

How well does your calendar agree with the local records kept by the Botanic Gardens or Park Department? Which is the most common tree in the neighborhood? How does altitude make a difference in time for the opening of buds? Go up a big hill or a nearby mountain.

Make a similar calendar for the first appearance of fall colors in trees. Local residents, tourists, and perhaps newspapers would be very interested in seeing this. Which color appears first? Which trees drop their leaves first? Are these the same trees which started earliest in the spring?

Try your skill at making simple calendars for the most common garden and wild flowers. These schedules can also be prepared by young scientists who live in the city. There are always parks, gardens, and small fields. Plants grow everywhere!

Will frozen seeds sprout?

If seeds are kept in the freezing compartment of a refrigerator, will they still be able to grow if they are later given the proper conditions of warmth and other necessities of life? Does it make any difference how low the temperature is, or how long the seeds are in the freezer? Can seeds of all kinds be equally frozen and survive?

This research can lead you along many paths. But perhaps you should first review some facts about seeds. A seed is one of the ways that a plant can continue the life of its species. Each seed contains a tiny, living baby plant called the EMBRYO (EM-bree-oh). There is also some stored food for the baby, and for the young plant when it first sprouts and has not yet produced its own food. The start of growth is called GERMINATION (jermin-AY-shun), and you can see this when the root begins

to grow out of the seed. The young sprouting plant is a
SEEDLING.

The baby plant and stored food are both covered by a
seed coat. This can be quite hard and tough in some
seeds, and it is capable of preventing injury as the seed is
moved in the soil. It also does not allow water to penetrate easily.

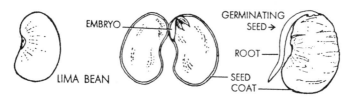

EMBRYO

LIMA BEAN

GERMINATING
SEED →

ROOT

SEED
COAT

Many plants release their seeds in the late summer. If
each seed started to grow at this time, the young plant
would be killed by the coming frosty weather in the
northern latitudes. Instead, the seed remains DORMANT
(DORE-ment) until the next growing season in the spring.
The dormant period is a resting time, a kind of extremely
slow living. The embryo is not dead. Indeed, some
dormant seeds can remain in this condition for many
years. The dried lima beans in your mother's pantry can
stay dormant for many seasons. Then if you plant them,
they will germinate.

Your experiments will show how much you can freeze
certain seeds. Put the seeds into a paper envelope. You
can work with dried lima beans, peas, seeds of sunflower,
radish, clover, flax, grass, and others. Always use *many*
seeds when doing each experiment. If you only used
one seed and that one died, people could rightfully say

that you used a dead seed or a diseased seed to start with.

Whenever you put an experimental group of seeds into the freezer, always keep a similar group outside the freezer as your control. Later you will see how many frozen seeds can germinate. At the same time, under the same germinating conditions, you will compare results with your control group.

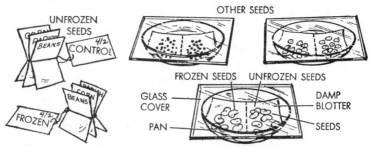

There are several good ways to test whether seeds will germinate. One way is to find a suitable dish and place white blotting paper, paper toweling, or some sawdust on the bottom of it. Keep the bottom *moist*—but not flooded. Flooding kills embryos in seeds by robbing them of air. Place your seeds to be tested in the container and keep it covered with a clear glass dish. This will help maintain a moist atmosphere. Keep at room temperature.

The seeds to be used as your controls should be kept in the dark before the germinating tests. This is because the experimental seeds were in the dark freezer. Many people allow all the seeds to soak in water overnight. This softens the seed coat and hastens germination.

Almost all refrigerators have thermometers in the

freezing section. When recording your results indicate the temperature, number of days or weeks the seeds were kept in the freezer, the kind and number of seeds, how many frozen ones germinated, and how many unfrozen control seeds germinated.

Believe it or not, foresters have discovered from such experiments that certain seeds which are exposed to very cold winters actually have a higher percentage of germination! Seeds of the yellow poplar will do this.

Just for fun, keep a group in back of the freezer for as long as you can. Even forget about it for a year. Do you think the seeds will germinate after such an exposure to cold?

Is the dormancy period a necessity for most seeds? Or can they grow immediately after they fall off the plant? Test whether fresh seeds will germinate. Many seeds need a rest period for "ripening." During this dormancy, certain chemicals called HORMONES are produced. These are necessary for germination.

You might also try freezing bulbs such as tulips, hyacinths, daffodils, and others, to see how they withstand cold. Also try potatoes and onions, and see what happens when they are kept below freezing for a period. See if they sprout when you set them halfway in a jar of water for several weeks. Have the same conditions for your controls.

Another idea to explore is this: Once the seeds sprout, can the seedlings stand being frozen? Start with similar seedlings and keep one batch in the freezer. Keep the control group at room temperature, but in the dark.

How short an exposure to the cold is necessary to kill seedlings? Can any survive?

What happens to many plants in the fields when there is an early, warm spring followed by a freakish spell of freezing weather? How can a farmer lose his investment in seeds if he plants too early? Look at a package of seeds. There is usually a warning: "Plant only when all danger of frost is over."

What affects the browning of apples?

It is a nuisance to put down a cut or partly eaten apple and have it turn brown. Your mother is annoyed too when her sliced apples turn "rusty" in her fruit salads. A little experimentation will help you know more about this.

Browning occurs when oxygen reacts with some chemicals in an apple. It is an example of slow oxidation,

WAYS TO COVER APPLE SLICES

APPLE CUT IN EQUAL PIECES

HONEY

CONTROL

OIL

BUTTER

GLASS JAR

WATER

WAX

WAXED PAPER

something like the rusting of iron. In the fruit there are certain substances called ENZYMES (EN-zimes). These hasten the chemical union with oxygen. When an apple is cut or bruised, the enzymes are released from the cells. Enzymes are destroyed by high heat and certain chemicals.

There are many varieties of apples on the market. Which brown most easily? Experiment by exposing slices to the air and recording how long it takes them to turn a certain brown color you decide upon. Does age, size, juiciness, or what part of the apple is used make a difference? Does anything in the weather influence browning; temperature, humidity, etc.?

Will there be less browning if you prevent air from getting to the sliced-apple surface? Cover slices from the *same* apple with wax paper, wax, jars, butter, honey, salad oil, or anything you can think of.

What chemicals hold back or prevent the reaction? Try covering some slices with citrus fruit juices, especially lemon juice. Also experiment with vinegar, vita-

TREATING APPLE SLICES WITH CHEMICALS

APPLE CUT IN EQUAL PIECES

OBSERVE EVERY 15 MINUTES

CONTROL

SALT

SUGAR

CITRUS FRUIT JUICE

Vinegar

VITAMIN C

min C solution, salt, and other harmless food materials.

In all the trials, you must expose to the air an untreated slice or two from the same apple, so as to have a control for comparison. Check your results at fifteen minute intervals, up to several hours.

You can also find out whether high temperatures prevent browning. Among other effects of heat, enzymes are destroyed. Cut several slices from the same apple. Boil one in hot water for a short time. Put another uncovered slice in the refrigerator and another slice in the open air. You might also try exposing a slice to a high oven temperature—but not long enough to brown it. How long can a slice which has been cooked remain exposed without turning brown?

Bruise an apple by striking it without cutting the surface. See if browning occurs at the damage spot. Allow an unbruised apple to dry up and wither away with old age. Does it show any browning? Do green, unripe apples get brown if they are sliced?

Do pears, bananas, and other foods also get brown? Make a list of foods which do. Don't forget the potato.

Ask your grocer for any of the preparations on the market for holding back browning of apples, especially for fruit salads. Some bottles may list the ingredients. One company claims it uses ascorbic acid, which is vitamin C. Do your experiments agree with this? What recommendations can you make to retard browning, after doing all your research?

ANIMAL STUDIES

What can change the respiration rate of a fish?

All living plants and animals need oxygen. They must also get rid of carbon dioxide, which is one of the waste products. Nature has many systems for doing this. For example, people have lungs for inhaling and exhaling. So do many animals. This process is called breathing and it allows air to get into the lungs. Here the air gets as close as it possibly can to millions of microscopic blood vessels called CAPILLARIES (CAP-ill-eries). These have very thin walls through which the different gases can pass.

The word "breathing" is used by biologists, usually, for animals having lungs. When a fish takes in oxygen and gives off carbon dioxide, the process is called RESPIRATION (res-pih-RAY-shon). Instead of lungs, a fish has four or five gills on each side of its head. These contain thin blood vessels through which can pass the invisible air which is dissolved in the water.

You can see the air that the fish takes in if you let a glass of water stand in a warm place. After a while there will appear tiny bubbles, which are driven out of the solution by the warmth. These are air bubbles and one-fifth of each bubble is oxygen.

Watch a goldfish in an aquarium. It seems to be swallowing water. Actually, it is forcing water over its gills. See how the gill covers open and close in rhythm with its mouth movements. When the fish opens its mouth, water rushes in. Then the mouth is closed. Now when the fish raises the bottom of its mouth slightly, the water is forced out over the gill openings.

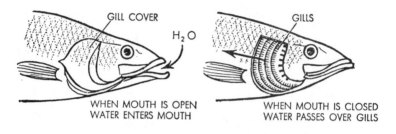

WHEN MOUTH IS OPEN
WATER ENTERS MOUTH

WHEN MOUTH IS CLOSED
WATER PASSES OVER GILLS

The empty mouth is filled again when the fish opens it. Each time the process is repeated, the bony gill covers open and close. In this experiment, let us call the number of times per minute that the gill covers open the RESPIRATION RATE.

When people are active they breathe faster. This is because their muscles and other body cells need more oxygen. They must also get rid of carbon dioxide. Will the respiration rate of a goldfish increase if you arouse the fish?

For best visibility, use a tank with flat sides. Start with a goldfish that is rested and has been very quiet for some time. Approach the aquarium slowly so as not to frighten the fish. Count the number of times the gill covers open in one minute. Count it several times to get the average, and rule out errors. If possible use a kitchen, one-minute interval timer.

Now agitate the fish by worrying it with a pencil. Chase it all over the tank, first a little, then faster, and finally do it very actively. Take its respiration rate each time. Does the rate increase as much as yours does when you run up the stairs many times? Try this experiment on many different fish. Is there much of a variation among them in normal respiration rates and in agitated rates? Take notes as follows:

	RESPIRATION RATES OF GOLDFISH DURING EXCITEMENT AT 72° F					
LEVEL OF EXCITEMENT	TESTS OF DIFFERENT FISH					
	No. 1	No. 2	No. 3	No. 4	No. 5	AVERAGE
At rest						
slight						
moderate						
much						

In order to conclude that the respiration rate is increased by more activity, you must be sure that the temperature is the same at all times. So remember to take the temperature of the water during every trial. You may have heated the water with your warm pencil, fingers, and breath. Temperature has a tremendous influence upon respiration rate, as you will learn in your next research experiment.

Fish are cold-blooded animals. Their body temperature is not maintained at an almost even level like ours.

111

Instead, they become as warm or as cold as their surroundings. When they warm up, the action of the body cells increases. This means that they need more oxygen. On the other hand, most cold-blooded animals can stand very cold temperatures. Their activity gets much less and, of course, their need for oxygen is decreased also. In the following experiments you are going to find out the changes in respiration rate. You will test goldfish who will be subjected to water temperatures from close to freezing up to about 90 degrees Fahrenheit.

This time start off with just enough water to completely cover the fish. Record the temperature and the number of times a minute the gill covers open. For your second reading, add enough shaved or crushed ice to the water to bring the temperature down about 5 degrees. Experience will enable you to get temperature changes in equal steps of 5 degrees. However, a degree or two difference once in a while is not important, as long as it is reported correctly.

Add the ice slowly so you do not shock or frighten the fish. Gently swish the water once, then wait a while. The water temperature should be the same all over. Continue lowering the temperature 5 degrees and taking the respiration rate. As you approach the freezing point (32 degrees Fahrenheit), you will notice hardly any movement of the gill covers.

To get the water above room temperature, add warm water very slowly. To maintain the water level at each trial, you can take out as much water as you add. Again do this in 5-degree stages. As you approach 90 degrees you will have to count very fast. The fish may gulp for air at the surface, because it needs more oxygen now. The fish may also be gasping because the warm water has less air dissolved in it than before. Put the fish back into cooler water as soon as possible.

Do these tests on as many goldfish as you can get. Do not use tropical fish. They are not as hardy as goldfish, especially when the temperature is lowered. It is also not as easy to observe gill movements. Many people find Comet goldfish best.

First make all your notes on a chart like the one below. Get the average respiration rate from all your fish for each temperature and then make a graph.

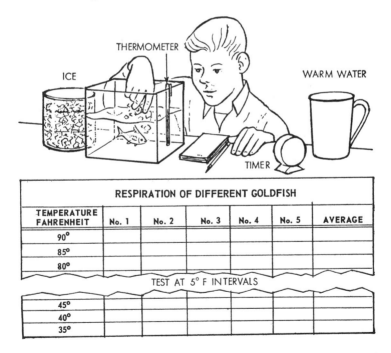

RESPIRATION OF DIFFERENT GOLDFISH						
TEMPERATURE FAHRENHEIT	No. 1	No. 2	No. 3	No. 4	No. 5	AVERAGE
90°						
85°						
80°						

TEST AT 5° F INTERVALS

45°						
40°						
35°						

AVERAGE
RESPIRATION
RATE OF
GOLDFISH

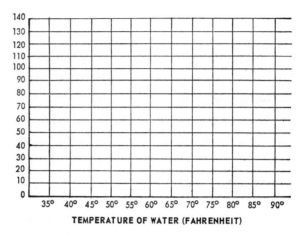

TEMPERATURE OF WATER (FAHRENHEIT)

Can you think of other experiments to change the respiration rate of goldfish? Will different amounts of aspirin, salt, sugar, bits of cigarettes (nicotine), music, or other vibrations influence the rate?

Can salt-water fish get used to fresh water?

Have you ever taken a walk along the inlets, ditches, and salt marshes near the ocean? You can see thousands of tiny salt-water fish shimmering and darting in the shallow water. A great many of these are killifish. Fishermen often use them for bait. Children sometimes take them home to be kept as pets in aquariums.

Can these small salt-water fish survive in tanks of fresh water? Will more survive if they are first placed in salt water which is then gradually made less and less salty? Will they finally become so accustomed to fresh water that they can be kept with goldfish or tropical fish?

You can easily catch many of these small fish in places near the ocean or at the edges of shallow marshes. Use nets, screens, or a seine. At the outset, place them in tanks containing actual sea water. Get the water from uncontaminated places and filter out any bits of dead material. It is good to use a small aerator in marine aquariums. Changing some of the water from time to time is another way to keep the water fresh. Salt water is very corrosive to metals, so use glass aquariums.

If it is too much trouble to keep getting sea water you can buy an inexpensive substitute. Use the special salts for marine aquariums which are sold at tropical fish shops.

The killifish and other species which you have caught have been adjusted to live in salt water. In your experiment you will remove daily a glassful of salt water and add an equal volume of fresh water. Keep observing

the fish and recording your dilutions. After a while, your sea water will become very dilute. In fact, it will almost be like fresh water.

In your final test, transfer the fish from this experimental tank to a tank of fresh water. Be sure the temperature is the same. You should find that it is possible for certain small salt-water fish, especially killifish, to get accustomed to fresh water.

You have enabled them to make this adjustment slowly. What would happen if these fish were taken from sea water and placed in fresh water immediately? Try this. What conclusions can you draw?

When doing experiments of this kind it is most important to have proper controls for comparison. Your control tank should have the same number of fish of the same size, kind, and condition. The feeding for both groups should be identical. The only difference is that the control tank has the sea water in it. In this way, you can see in what condition the fish would be if they were in their original salt-water environment. Otherwise, if the fish died in fresh water you would never be sure if the fish were just weak, and would have died anyway in the same length of time.

You might wish to do another experiment dealing with salt water and fish, but this time in reverse. In other words, can a guppy, which is a fresh-water fish, get used to sea water? If your hobby is tropical fish, you usually have enough guppies for this experiment. Guppies breed quickly and in great numbers.

Before you begin, you may already know that most tropical fish can live in water which is somewhat salty.

When these fish are sick, the recommended treatment is to add one tablespoonful of table salt to each gallon of water in the aquarium. This water need not be changed when the fish recover from their illness. However, sea water has much more salt in it than this.

Start off with two equal containers of suitable fresh water for the guppies. Into each container place the same number of guppies of the same kind, size, and condition. Keep one tank as a control. The experimental one should have the same temperature, light, and food. The only difference between the two tanks will be in the salt content, which you will change as follows:

At certain intervals, which you will determine, you will scoop up some water in a glass from the experimental tank. Dissolve a certain amount of salt in it, say a teaspoonful. Pour back the glass of salt water evenly, all over the aquarium.

If possible, add the salt used for marine tanks. If you cannot obtain this, table salt may be substituted. Record the dates and the amounts of salt you add. Also write down the condition of the fish.

Can a guppy slowly get accustomed to salty water? Can it live in the sea water?

Can certain animals be frozen and survive?

Some people are firmly convinced that some day there are going to be space stations hundreds of miles up in the sky. Astronauts will use these as springboards to destinations which may take 100 years to be reached. No ship can possibly take along enough air, food, and water for such a trip. Some experts believe that there is a solution to the problem. They say that the men will have to be placed in a state of suspended animation by greatly lowering their body temperatures.

In this almost frozen condition there will hardly be any body activity or the need for the necessities of life.

Years later at a predetermined time, an electronic computer will thaw out the 125-year-old, or rather young, men. They will then go about their business as though they had just awakened from a peaceful night's slumber.

At present, this is right out of the imaginative brain of a science-fiction writer. However, there is some basis for this kind of reasoning. More and more, doctors and biologists are developing techniques for keeping people alive, though extremely inactive, under reduced temperatures. The science is called HYPOTHERMIA (high-poh-THERM-ee-uh). Many operations, especially those on the heart, are done while the body is cooled by ice and cold water. Hypothermia specialists say that a person's temperature can be safely lowered to about 80 degrees Fahrenheit for short periods.

As you learned in the experiment, "Will Frozen Seeds Sprout," on page 102, many seeds are constantly being frozen and yet they survive. Certain cold-blooded animals can remain alive while buried in frozen mud. Oysters can live in surroundings which are below the freezing point of water. The pupas of butterflies and moths

118

and also the eggs of insects are often frozen in the wintertime.

Here are some experiments you can try in the family's refrigerator. If your mother dislikes the entire idea, convince her that it is all done in jars. It is sanitary, neat, and interesting. If all else fails—tell her it is very educational. No parent can resist such an argument!

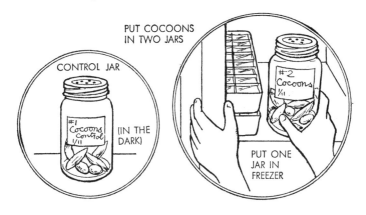

Find many cocoons of different moths. Discard those which seem to feel completely empty. Divide the good ones into two equal groups containing the same number and the same kind. Put each group into a separate covered jar. Make a few holes in each cover for ventilation. Keep one jar in the freezer and the other at room temperature. Label each one. The unfrozen jar is your control. After a certain period, remove the jar from the freezer. At the proper time and under suitable conditions, see which cocoons will produce perfect moths.

Record the cold temperature and the length of time that the cocoons are in the freezer. Try different tem-

119

peratures. The time too can vary from one hour to an entire winter. When a jar is removed from the cold, open its cover and also the cover of its control. Once in a while, sprinkle the cocoons in both jars with a *few* drops of water. Do not soak them. The drops will prevent the cocoons from drying out. Too much water may turn the cocoons moldy.

If you wish, you may remove the cocoons from both the control jar and from the one in the freezer. Put them into a large open container so that they are not crowded. Make sure each cocoon is identified in some way.

Try an experiment to see whether earthworms can be frozen for certain lengths of time. Always handle containers carefully when removing them from the cold because frozen animals are very fragile. All thawing must be done *slowly* at room temperature. Never use hot water or other heating devices.

Do the same with beetles. These are amazingly hardy. Use controls. Also experiment with flies, mosquitoes, grasshoppers, and others. Do you think caterpillars can be successfully frozen and thawed out? Feed your control in this case. For food, use leaves on which you found the insect.

TRY FREEZING OTHER ANIMALS

Can cold (not necessarily freezing) hold back the development of a caterpillar for any period? And will there be any injury to the butterfly or moth which it will form?

You may sometimes read about experiments where scientists have frozen goldfish in cakes of ice and successfully thawed out the fish. Some people do not seem surprised at this. They say that fish are always frozen solidly into the ice when a pond or lake freezes.

However, most of the time the fish are in the water *under* the ice. Ice expands when it freezes. This makes it lighter than the water and so it floats on top of the pond. The fish are quite inactive because of the cold. They hardly need any food. There is always air under the ice. In this way, the fish survive the winter.

Of course, very shallow ponds or lakes may freeze solid during a spell of extremely frigid weather. Yet, when spring rolls around, you may see fish swimming in the water. Since nobody put them there, one can assume that the fish or the eggs were frozen and survived.

Will a moth develop without a cocoon?

With sharp scissors cut open a large moth's cocoon. Do not harm the pupa inside. Hold the insect for a while in the warm palm of your hand. You will see the pupa stir and often make strong twitching movements.

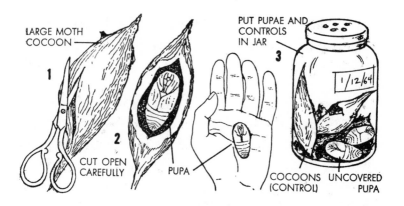

LARGE MOTH COCOON

1

2 CUT OPEN CAREFULLY

PUPA

PUT PUPAE AND CONTROLS IN JAR

3

1/12/64

COCOONS (CONTROL)

UNCOVERED PUPA

In the cocoon the pupa will develop into a perfect moth. The cocoon will supply it with warmth, humidity, and protection from many unfavorable outside conditions. Can the pupa without its natural home still successfully finish its transformation?

Try to develop pupas which have had their silky cocoons removed. Place naked pupas in a jar. Next to these lay uncut cocoons containing similar pupas. These are your controls. Close the jar with a cover which has been pierced with some holes. Keep at room temperature. Sprinkle a few drops of water into the jar once a week.

Can you think of any kind of covering you can put over other naked pupas to serve as useful cocoons? Plastic, cloth, wax paper, aluminum foil, felt, absorbent cotton, Fiberglas, insulation, etc.?

Remember to leave enough of an opening at one end of the covering for the emerging moth to escape.

Are hummingbirds attracted to color?

Hummingbirds are like helicopters. They can hover in the air, fly up, down, forward, and even backward! They are very interesting to watch as they dip their remarkably long bills into the centers of cuplike flowers, such as gladiolas. They sip the nectar this way.

You can easily attract hummingbirds if any are in the neighborhood. Obtain some narrow glass or plastic pill containers. Pour into each one some honey, or a syrup made of sugar and water. Tie a red ribbon around the neck of each small bottle. Do not have a ribbon on one bottle which will serve as a control.

Suspend these from low branches, stakes in the ground, or from any other support. Do this in an open spot where you can observe the birds as they come to feed. Do not keep these bird feeders too close to where you will

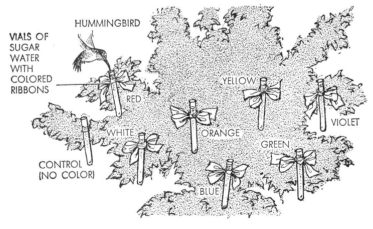

VIALS OF
SUGAR
WATER
WITH
COLORED
RIBBONS

HUMMINGBIRD

RED

WHITE

CONTROL
(NO COLOR)

YELLOW

ORANGE

GREEN

BLUE

VIOLET

be observing. Stay perfectly still. Fast motions frighten birds.

To find out if hummingbirds are attracted to other colors besides red, place several similar containers in a cluster. Tie a different, bright-colored ribbon near the top of each. Keep records of hummingbird visits and which colored container seems to be preferred. Ask your friends to help you. Keep a tally sheet handy for their observations.

ABOUT YOURSELF

How accurately can you hear
under noisy conditions?

In spite of all modern devices used in communication, there is usually some point where people must have direct contact in order to deliver a message. Words are exchanged and these are often received incorrectly. There are many reasons for error. But in the following experiments you are only going to consider noise.

Noise is an old problem faced by many individuals, businesses, and organizations. For instance, all the armed forces recognize the importance of receiving messages very accurately. These are heard under noisy battle conditions and through radios crackling with static.

Much research has been done on this problem and attempts have been made to avoid messages which use the voice. However, where voice contacts must be made, a system has been devised for checking important words. These are spelled out, with words used instead of letters. You surely must have heard these used in some war moving pictures.

These words have been found to be more easily understood than letters. Almost all of them have two syllables and distinctive sounds. If only part of the word is heard, the rest can be guessed fairly accurately. Here is the list used by all the armed forces for the letters of the alphabet:

Alfa	Juliet	Sierra
Bravo	Kilo	Tango
Charlie	Lima	Uniform
Delta	Mike	Victor
Echo	November	Whiskey
Fox trot	Oscar	X-ray
Golf	Papa	Yankee
Hotel	Quebec	Zoo
India	Romeo	

Suppose the word to be transmitted is "east." To prevent error it may be checked out as Echo Alfa Sierra Tango. The telephone company and people who do much business by telephone also have their own list of key words.

In your research you will wish to find out which numbers, letters, vowels, or words are more easily misunderstood. Do boys or girls make more mistakes? Do male or female voices sound clearest through noise? Does more noise cause more confusion for the careful person as well as for people in general? What recommendations can you make for word substitutions?

You will need some method of getting noise which will be similar for all your experiments. There must also be some way to make the noise louder and to be able to note how much louder. One idea is to use your record player. Put on an old record with fast, "jumpy" music. Enlarge the hole so it is off center. Play it at the wrong speed; that is, use a 78 speed for a 45 or 33⅓ rpm. This should provide good background noise.

An excellent noise can be obtained by having two record players working at the same time. If you have a tape recorder, you and your friends can have a wild, hilarious time recording a noisy background which will drive the devil himself insane.

Make reference marks on the volume control for "low," "medium" and "high." In this way, you can always reset the noise level to the same reading for different experimental subjects.

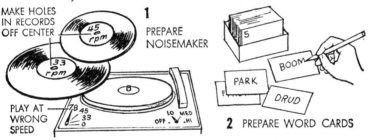

MAKE HOLES IN RECORDS OFF CENTER

PLAY AT WRONG SPEED

1 PREPARE NOISEMAKER

2 PREPARE WORD CARDS

Get 100 pieces of paper or library cards and on each write a different number from 1 to 100. You may use the back of each card to write the words you will use. The cards should be thoroughly shuffled when used for a subject who is receiving the same words, but at different noise levels.

Prepare a list of words which uses every letter and vowel many times. It is best to use one-syllable words because a person may be able to guess a word having many syllables if he only hears part of it.

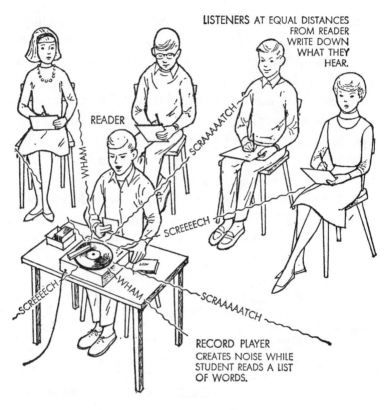

LISTENERS AT EQUAL DISTANCES FROM READER WRITE DOWN WHAT THEY HEAR.

READER

SCRAAAAATCH

SCREEEECH

WHAM

SCREEEECH

WHAM

SCRAAAAATCH

RECORD PLAYER CREATES NOISE WHILE STUDENT READS A LIST OF WORDS.

On some future tests you might also try making up nonsense syllable words such as rell, drud, koom, mork. Do not mind the spelling. Count the word correct if the sound is accurate. This type of technique rules out guessing and is used by expert investigators.

In giving the test, the caller (sender) of the message sits with his back toward the person who is receiving the message. Any number of people may receive at the same time as long as they sit the same distance away from the person who is reading from the cards. Of course, the distance from the record player must also be the same for all.

The caller must speak the words in the same way all the time. This is the most unscientific part of this test, since you have no way of being certain that the caller's voice is always at the same volume or equally clear. But since you do not have the expensive facilities needed to correct this, you will just have to ask the sender to be very careful. Can you think of any other scheme? One of your friends may have a tape recorder. Then you could make a series of tape recordings needed for all the tests. It is something to consider.

Correct the test sheet for each person. On a master sheet, tally the number of times each test word was received incorrectly. Record the total number of errors made at different noise levels. Keep accurate records of other results. If you make good charts and graphs, you will be able to arrive at your conclusions more easily.

Here are some samples:

	TOTAL ERRORS RECEIVED		
WORD CALLED	VOLUME OF NOISE		
	LOW	MEDIUM	HIGH
ant	١١.	١١١	₩ ١
breath	١١١	₩ ١١	₩ ₩ ₩ ₩
stamp	١	١١	١١١
dark	₩	₩ ١	₩ ١١١
grunt	١١	١١١	₩ ١١
kings	₩ ١١	₩ ₩	₩ ₩ ١١
first	₩ ₩	₩ ₩ ₩ ₩	₩ ₩ ₩ ₩ ١١

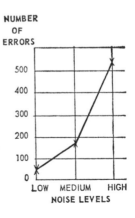

NUMBER OF ERRORS

500
400
300
200
100
0

LOW   MEDIUM   HIGH
NOISE LEVELS

MEASURE
HEIGHT OF STEP

MULTIPLY BY
NUMBER OF STEPS

RECORD THE
TIME FOR
RUNNING
UPSTAIRS

What is your horsepower?

To a scientist, the word "power" means the rate of doing work. James Watt, who is famous for developing the steam engine, wished to express power in definite units. In his day, horses were used for much work—so he used a horse as his basis for comparison.

He found by actual experiment that a horse can pull or lift 550 pounds, 1 foot high, in one second. He called this rate of working ONE HORSEPOWER. The abbreviation is h.p.

Today horses are rarely seen in cities, but still this unit of power is very much with us. When your father buys a car, his engine's rating may be about 250 h.p. All the electric motors in your home are in horsepower sizes.

Look for the h.p. on the manufacturer's plate on each motor. Airplane engines, diesel locomotives, and even air conditioners are rated in horsepower.

It has been found that a man can lift 55 pounds, 1 foot high, in one second. His rate of doing work would be $\frac{1}{10}$ h.p. Actually an athlete can exert, in short spurts, even more than 1 horsepower. But he cannot keep it up for more than a few seconds.

It should be interesting to find out how much horsepower you can develop. First find a long flight of stairs. Make sure that the stairs are not broken or slippery. Measure the height of one step. Count how many steps there are and multiply this number by the height of each step. This product will give you the *vertical* distance you are going to lift yourself. Your answer must be in feet. If possible, you can use a string with a weight at one end for measuring the vertical distance from the top step to the landing.

Now get a friend to help you. He needs a watch with a good second hand. Perhaps you can borrow a stopwatch from a teacher. Have the timer see how long it takes you to reach the top of the stairs. You may take more than one step at a time. You may use the railing. You may even fly if you can!

Use the following formula to learn how your rate of doing work compares scientifically with that of James Watt's horse.

$$\text{h.p.} = \frac{\text{number of vertical feet} \times \text{your weight}}{550 \times \text{number of seconds}}$$

For example, suppose the vertical height from the

landing to the top step is 33 feet. Your weight is 100 pounds. The time is ten seconds. Substituting in the formula, we get:

$$\text{h.p.} = \frac{33 \times 100}{550 \times 10} = \frac{3300}{5500} = \frac{3}{5} \text{ h.p.}$$

How do eyes help you keep your balance?

In our everyday contact with modern electronic miracles, we often forget what remarkable mechanisms our own bodies contain. For example, our sense of balance depends upon amazing team work of many parts of the body.

We stand erect because muscles all over our body are constantly correcting our upright position. If you think about it, you can almost feel your foot, leg, and thigh muscles relaxing and tightening. Doctors say that muscles can "sense" an unbalance and automatically adjust our posture. Learning to walk as a child consisted partly in training these muscles.

The main balancing organs are in the nonhearing part of our ears. In each ear we have a SEMICIRCULAR CANAL. This consists of three looped tubes, each at right angles to the other two. These loops are filled with fluid and lined with nerve cells going to the brain. A movement of the body in any direction causes the proper muscles to move to keep us balanced.

In each ear we also have tiny, loose, solid particles.

HOW LONG CAN YOU STAND ON ONE FOOT
WITH YOUR EYES CLOSED?

When the head is tilted, gravity causes these to touch ends of nerves leading to the brain.

But you probably do not realize how important your eyes are for balancing until you do the following experiment. Keep your eyes open and stand on one foot. Unless you are a nervous wreck, this is fairly easy to do.

Now close your eyes and stand on one foot. See how many seconds you can do this apparently simple stunt. Suddenly you realize that there is something wrong. Without your eyes to give you clues, all your other priceless balancing machinery in your ears and brain seem to be out of order. Yet as soon as you open your eyes you are again master of the situation.

Try many schemes to help you keep your balance on one foot with your eyes closed. Before you close your eyes look closely at a strong light. Does the lingering light impression in the eye help you in any way? Try bending the leg you are standing on to lower your center of gravity. Twist your foot or your body to find a more stable one-legged position. Orient your position by standing close to a source of sound. These ideas may not work, but they are the kind of experimentation you should try.

Can you train yourself, after a while, to stand erect on one foot? Do you get the same swaying effect when you stand for several minutes on both feet in the dark? Do tightly laced shoes help? Try this balancing stunt in the early morning or late at night. Does it make any difference whether you are tired or fresh?

Keep records of the number of seconds you can count off each time. Do you find that certain people can do this better than others? Ask these people if they ever get dizzy, seasick, or carsick. It may be that people who can stand on one leg with their eyes shut have more stable nervous systems as far as balance is concerned.

Can you see blood vessels in your own eyes?

Have you ever seen a doctor look into a patient's eye with an instrument called an OPHTHALMOSCOPE (ahf-THAL-muh-scope)? The device lights up and magnifies the *inside* of the eye, especially the meshwork of blood

134

vessels of the retina in the rear of the eye. This gives the physician a clue to the general condition of the rest of the body's arteries, veins, and the tiniest blood vessels called capillaries.

This experiment is not referring to the tiny blood vessels you may easily see on the white surface of your eyes. There is a way for you to see the blood vessels in the rear of your own eyes—and without the ophthalmoscope. All you need is a flashlight and a very dark room. Cup your left palm over your left eye. With your head erect, keep your right eye open and look toward the floor. Now shine the flashlight toward the ceiling, keeping it a little below eye level. Do not look directly at the beam, but keep looking at the floor. Vary the distance between the flashlight and your face.

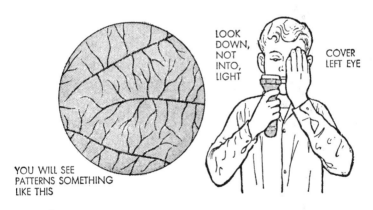

YOU WILL SEE PATTERNS SOMETHING LIKE THIS

LOOK DOWN, NOT INTO, LIGHT

COVER LEFT EYE

Move the flashlight at slightly different angles, but always toward the ceiling. At certain positions, you should now see a network of black blood vessels on a faintly orange background. The blood vessels resemble jagged lightning bolts or trees in the wintertime. The crooked

135

lines do not stand still but keep changing all the time.

The image appears to be in front of the body and quite enlarged. It takes a little practice to see this. Sometimes your eye gets tired before you are successful and you may have to use the other eye or take a rest. Sometimes, if you suddenly change the direction of your eyes from left to right, you will see the blood vessels very brightly for an instant.

The explanation of this effect is as follows: The retina is kept alive and healthy by a meshwork of blood vessels over it. The flashlight beam coming at a slant makes these blood vessels cast their shadows on the light-sensitive retina underneath.

Normally there are always such shadows on the retina, but we have learned to disregard them. But in our experiment we purposely had the light enter the eye at a great slant. The shadows were then cast on unaccustomed places on the retina so that we became more aware of the shadows.

Can you make a drawing of what you see? Does it look like the drawings made by some of your friends?

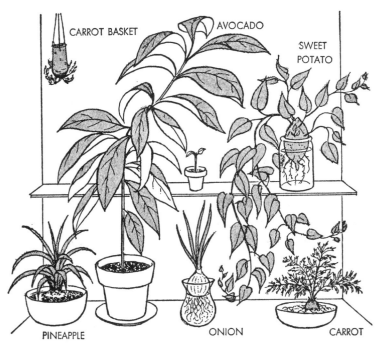

CARROT BASKET   AVOCADO

SWEET POTATO

PINEAPPLE   ONION   CARROT

SCIENCE IN THE HOME

Can you make an unusual window garden?

Do you envy people who have "green thumbs" and are able to make anything grow? They were not born that way, but acquired the knack from experience. Would you like to have a colorful, cheerful garden right in your own kitchen, even through the winter? You can easily make one which will catch the eyes of all your visitors and be a conversation piece. To make this unusual thing of beauty and wonder you must do some enjoyable research work.

In this experimental indoor garden you will use all kinds of common household vegetables and seeds. These are inexpensive and are very good natural producers of green leaves and vines. You will be fascinated by the different shapes and designs each plant produces.

Try to grow the tops of root vegetables such as carrots, beets, turnips, radishes, and all kinds of other roots you may find in the vegetable store. Cut off about an inch or two from the top of the root.

Place this top part upright in a deep dish containing water and small pebbles or fish-tank gravel. Wedge these small stones against the cutting, for support. The water must *never* completely cover the vegetable, which may spoil and perhaps give the water a bad odor. Part of the plant must always be out of the water in order to get air.

Add more water when necessary and in a week or two each vegetable will produce lacy, fernlike or reddish leaves. It is fun to see the variety. These leaves are not receiving any nourishment, as they would if they were in soil. So when all the food in the vegetable is used up, the leaves will wither away. Nevertheless, you will usually have many weeks of growth.

There is another kind of experimentation in your search for variety for your window garden. You can grow strange seedlings from all kinds of seeds found in your home. Place the seeds in a dish on a moist blotter, paper towel, absorbent cotton, sawdust, sponge, or other holders of water. Again, the seeds must *not* be under water—just moist. Place an inverted jar or glass over the seeds. This will produce good, hothouse humidity.

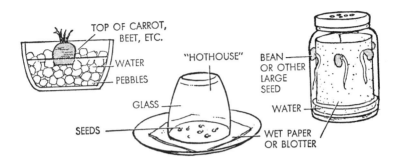

Try clusters of small seeds such as grass, bird seed, flaxseed (not ground), and also flower seeds used for outdoor gardens. You may also use large seeds such as lima and many other beans, corn, peas, and lentils.

If seeds of citrus fruits are used, wash them well. You can grow grapefruit, lemon, orange, and lime. Also experiment with seeds of apple, pear, watermelon, cucumber, tomato. Try pumpkin seeds and peanuts (both unroasted!), also acorns and maple keys. Soak most seeds completely under water overnight before using.

Some seeds taken from fresh vegetables may have to be dried first for several weeks. So, if you have no success with very fresh seeds, it is because they are not yet "ripe" for growing. If larger seeds do not germinate, try nicking them carefully with a knife or file. This helps the tough seed coat to open more easily when the moist, growing seedling pushes against it.

You can also grow larger seeds such as beans and corn by the following technique: Curl a long white blotter inside a glass. Place the seed halfway down the glass, between the blotter and the glass. It should stay there without falling. Add water so that it is only about 1 inch deep in the container. The blotter will get moist all the way up to the seed.

Do you know that a house brick is porous? Put one partway into water in a dish, and the top will become moist. Place grass or other small seeds on it and in a week or so you will have a novel attention-getter!

The seed of an avocado (alligator pear) makes an unusual treelike plant. First remove the hard coat, then set it over a narrow jar of water. It should be only partly in the water. If the seed is smaller than the jar opening, you might try to stick toothpicks into it to hold it in place. See illustration. Keep in a dim light, or even in the dark, for three weeks or longer, until the roots appear. Make sure there is always enough water to touch the seed.

A sweet potato or a yam produces vines with long, beautifully veined leaves. Wash well and then set into a glass or jar, with the pointed end down. It should be only partly in the water. Keep in a dark place until the roots develop. Wash off any slime or mold which may develop. Change the water if it has a bad odor.

Since the above plants are grown only in water, they are living on the food stored in the root, seed, or stem. When all the nourishment is used up, the new plant will die. But if you wish to prevent this, you can place any one of the new plants in soil.

The top of a pineapple produces good greenery too. Cut about one or two inches from the top and set it into water and pebbles.

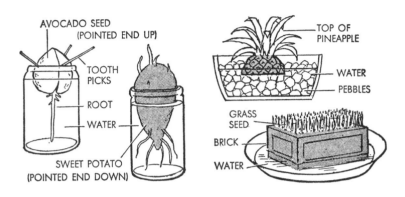

Experiment with ordinary potatoes, onions, hyacinths, tulips, narcissus, and anything you have not tried before. Go for a walk in the wintertime and bring back twigs containing buds of pussy willows, cherry, apple, forsythia, and dozens of others you are curious about. Put the ends in water and you can have springtime in January right in your kitchen!

Make a hanging basket from a carrot or other root. Cut about 4 inches from the broad end. Carefully scoop out the narrow end so that the carrot is like a deep cup. Tie string or copper wire around the carrot so that it can be suspended. The former top of the carrot is now on the bottom. See illustration on page 137. Keep the hole filled with water. You will be rewarded with a very pretty hanging plant.

These plants do not need much care. But they do demand certain necessities. As you know, they need a certain amount of water. Most of them also need sunlight. Another requirement is that the temperature should be moderate. Beware of the hot drafts over radiators or the freezing blasts of air from openings around windows.

Look for a good location for your experiments. Some windows can have shelves put across them. Perhaps a properly protected table near a window would help keep your experimental indoor garden in the public eye. Your parents will eagerly help you in this interesting and educational project.

Will any of your plants last longer if a tiny pinch of plant fertilizer is added to the water? Do an experiment and use a control. You might also try to find out if it is beneficial for the plants to receive an occasional spraying from a watering can or laundry sprinkler.

The above suggestions are mainly for unusual ways of growing things. Meanwhile you can grow the usual household potted plants everybody knows about. Only by doing this yourself will you learn the necessary conditions for growth and the pitfalls to avoid.

Ask questions of people you know who seem to have remarkable success with house plants. Visit botanic gardens. Read the many books on plant care and how to make new plants from cuttings. You will develop a hobby that will remain with you for the rest of your life. But, above all, you will one day be delighted to hear others refer to you as "the one with the green thumb."

How many hours a day
is your TV set used?

Do your parents ever complain about the number of hours you spend looking at television during the day or week? How would you like to keep an accurate, *automatic* record of every second that the set is working?

There are many scientific experiments involving electric lights, heaters, motors, and other electric devices. It is often necessary to know how long these were in operation. The following simple, safe, and inexpensive technique will add another useful experimental device to your knowledge.

You will need a self-starting electric clock. You probably have several in your home. The idea is to plug the TV set and the electric clock into the same outlet. Then every time the set goes on, so does the clock, and it begins to register the elapsed time. There are several ways to do this.

Do you have a wall switch which controls an electric wall outlet? All you have to do is to plug both your TV and clock into this outlet. The wall switch puts the two devices on simultaneously. The outlet is usually a double one. If it is a single outlet, simply plug in a two-way adapter plug.

If you do not have such a handy switch found in many homes, here is another method. You surely must have some kind of a lamp near the television set. Unscrew the bulb, and replace it with a socket adapter having two

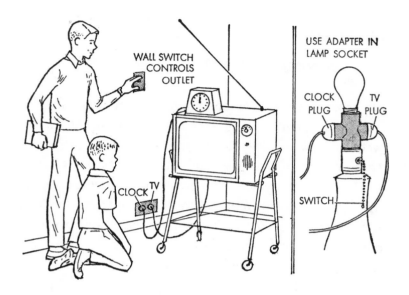

WALL SWITCH
CONTROLS
OUTLET

CLOCK TV

USE ADAPTER IN
LAMP SOCKET

CLOCK
PLUG

TV
PLUG

SWITCH

outlets on either side. Look at the one shown in the illustration. Your folks probably have this practical adapter available at home. Plug the TV and the clock into the adapter.

Now whenever the TV is to be used, the lamp switch must put it on. The clock will also go on at the same time. If you screw a small bulb into the adapter it can serve as a television light too. Of course, the switch on the TV set must be on all the time.

If you wish daily records, you must start the clock at twelve o'clock. Then take your reading after the last program of the day and reset it to twelve o'clock. If you do not take daily readings, remember that if the time of viewing runs beyond twelve hours, you must be aware of this fact.

Keep your records on a chart. It should also be interesting to make a graph of your family's passion for TV entertainment.

DATE		TIME SET WAS ON
Mon.	3/10	1 hour 15 minutes
Tues.	3/11	2 hours 40 minutes
Wed.	3/12	2 hours

NUMBER OF HOURS
TV SET WAS USED DURING
WEEK OF MARCH 10

If you wish, you may take several readings a day to find out when the set is used most during the day. It might be fun to check up on the viewing habits of your mom while everybody is at school or at work.

You might break up the day into three reports: up to 3 P.M., from 3 P.M. to 7 P.M., and from 7 P.M. to the last program seen at night.

Can you think of an easier way to set up the clocking of the programs? What other happenings might you time with this technique? For example, how many hours a month is the fish-tank light, heater, or aerator on? How long are the basement lights used? How much is the air conditioner used?

You can also figure the cost of operating any electrical

145

device. All you have to know is the wattage and number of hours the device is used. The wattage is on the bulbs and also printed somewhere on the apparatus. Find out the cost of a kilowatt hour (kwh) by looking at the back of your electric bill. This is the price you pay, in your community, for using 1,000 watts for one hour.

Substitute your information in the following formula. It will give you the cost of using any wattage for any number of hours.

$$\text{Cost} = \frac{\text{Watts} \times \text{Hours} \times \text{Cost of 1 kwh}}{1000}$$

For example: How much would it cost to operate a 200-watt TV set for ninety hours at 6 cents a kilowatt hour?

$$\text{Cost} = \frac{200 \times 90 \times .06}{1000}$$

$$= \$1.08$$

How does grass turn yellow?

Grass is green because of a coloring material called CHLOROPHYLL (KLO-ro-fill). This vital substance is used by the grass to manufacture sugar and starch from carbon dioxide and water. This process is called PHOTOSYNTHESIS (fo-toh-SIN-the-sis) and can be done only by green plants in sunlight.

Chlorophyll keeps breaking down because it is rather delicate, but it is replaced by the plant all summer, as fast as it is destroyed. It needs sunlight to do this. You can

146

prove this by placing on some grass an inverted box or cover which will block out the sun, but not the air or water.

How long does it take for grass to turn yellow? Use an edge of the lawn in back of your house which is not conspicuous. Lay several inverted boxes over the grass in a row. Pick one up every two days. Record your observations. Your control in this case is the surrounding grass. Is there a gradual yellowing all the way up to the area which was covered for fourteen days? Or does the amount of yellowing remain unchanged after a certain number of days? If necessary, continue the experiment for a longer period.

How long does it take for the yellowed grass to fully recover and look as green as the surrounding grass? Does it take just as long to get green as it did to get yellow? Does the green come back all over the blades of grass?

Or does the green come up from the ground, like the new hair of a woman who has had her hair bleached?

In all your recovery experiments keep records of the weather. You must consider the extent of direct sunlight shining on the yellow grass. Perhaps rain makes a difference too.

Will the grass generally lose its green color faster if it is kept dry, wet, closely cropped, or long? Cover all these prepared areas for a similar length of time. Inspect at frequent intervals. Again use the surrounding grass for your control.

Will all kinds of grass turn yellow and then green again at the same rate? Compare with somebody who has a different kind of grass on his lawn. Is it possible to keep grass covered so long that it never recovers because it is dead? What else besides lack of sunlight causes grass to lose its green color? Too much or too little fertilizer, beetles or other insects, draught? Why might any of these cause the yellowing?

If you place a heavy flat board or a brick on the grass, will the covered grass turn yellow faster than under an inverted box? Remember that in all the experiments the amount of water the grass receives must be the same.

Why is salt put on icy sidewalks?

You have often seen home owners throw table salt on ice and snow. In a short time the ice melts and the hard work of chopping and shoveling is avoided. You can learn easily about this by doing these experiments. You will need an outdoor thermometer because this kind has a scale which goes down below zero degrees. If possible, get some snow. Otherwise, chop some ice cubes in a cloth until they look like powder. Put it into a container and record the temperature. You will find it to be 32 degrees Fahrenheit.

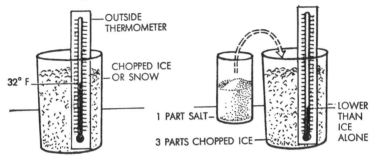

Now make a thorough mixture of one part table salt and three parts powdered ice. Stick the thermometer down deep into the ice and salt. This time you will observe the temperature falling rapidly. In fact, if you have a very good mixture, the lowest temperature you can get is minus 7.6 degrees Fahrenheit.

Salt lowers the freezing point of water. In other words, salt water freezes at a lower temperature than ordinary water. That is why salt melts ice on which it is placed. Research scientists have found that in order to have crystals of ice, the water molecules must arrange themselves into a certain pattern. But the salt does not allow this to happen. Sometimes another salt, calcium chloride, is used instead of table or rock salt. Ask your hardware dealer about this.

These salts are harmful to the lawn, concrete, rugs,

149

leather, and especially the chromium on a car. Can you design some experiments to prove this?

Will sugar or other soluble materials you may have at home act like salt and lower the freezing point of water? Try some.

You can make a very interesting graph by plotting the amount of salt you use against the temperature. Start off by *thoroughly mixing* one teaspoon of salt with the powdered ice. Allow the thermometer enough time to get the lowest reading. Record this temperature. Add another teaspoon of salt and repeat. Do the same with calcium chloride and other de-icers.

Index

151

152

A CATALOG OF SELECTED DOVER BOOKS IN ALL FIELDS OF INTEREST

DRAWINGS OF REMBRANDT, edited by Seymour Slive. Updated Lippmann, Hofstede de Groot edition, with definitive scholarly apparatus. All portraits, biblical sketches, landscapes, nudes. Oriental figures, classical studies, together with selection of work by followers. 550 illustrations. Total of 630pp. 9⅛ × 12¼.
21485-0, 21486-9 Pa., Two-vol. set $25.00

GHOST AND HORROR STORIES OF AMBROSE BIERCE, Ambrose Bierce. 24 tales vividly imagined, strangely prophetic, and decades ahead of their time in technical skill: "The Damned Thing," "An Inhabitant of Carcosa," "The Eyes of the Panther," "Moxon's Master," and 20 more. 199pp. 5⅜ × 8½. 20767-6 Pa. $3.95

ETHICAL WRITINGS OF MAIMONIDES, Maimonides. Most significant ethical works of great medieval sage, newly translated for utmost precision, readability. Laws Concerning Character Traits, Eight Chapters, more. 192pp. 5⅜ × 8½.
24522-5 Pa. $4.50

THE EXPLORATION OF THE COLORADO RIVER AND ITS CANYONS, J. W. Powell. Full text of Powell's 1,000-mile expedition down the fabled Colorado in 1869. Superb account of terrain, geology, vegetation, Indians, famine, mutiny, treacherous rapids, mighty canyons, during exploration of last unknown part of continental U.S. 400pp. 5⅜ × 8½. 20094-9 Pa. $6.95

HISTORY OF PHILOSOPHY, Julián Marías. Clearest one-volume history on the market. Every major philosopher and dozens of others, to Existentialism and later. 505pp. 5⅜ × 8½. 21739-6 Pa. $9.95

ALL ABOUT LIGHTNING, Martin A. Uman. Highly readable non-technical survey of nature and causes of lightning, thunderstorms, ball lightning, St. Elmo's Fire, much more. Illustrated. 192pp. 5⅜ × 8½. 25237-X Pa. $5.95

SAILING ALONE AROUND THE WORLD, Captain Joshua Slocum. First man to sail around the world, alone, in small boat. One of great feats of seamanship told in delightful manner. 67 illustrations. 294pp. 5⅜ × 8½. 20326-3 Pa. $4.95

LETTERS AND NOTES ON THE MANNERS, CUSTOMS AND CONDITIONS OF THE NORTH AMERICAN INDIANS, George Catlin. Classic account of life among Plains Indians: ceremonies, hunt, warfare, etc. 312 plates. 572pp. of text. 6⅛ × 9¼. 22118-0, 22119-9 Pa. Two-vol. set $15.90

ALASKA: The Harriman Expedition, 1899, John Burroughs, John Muir, et al. Informative, engrossing accounts of two-month, 9,000-mile expedition. Native peoples, wildlife, forests, geography, salmon industry, glaciers, more. Profusely illustrated. 240 black-and-white line drawings. 124 black-and-white photographs. 3 maps. Index. 576pp. 5⅜ × 8½. 25109-8 Pa. $11.95

THE BOOK OF BEASTS: Being a Translation from a Latin Bestiary of the Twelfth Century, T. H. White. Wonderful catalog real and fanciful beasts: manticore, griffin, phoenix, amphivius, jaculus, many more. White's witty erudite commentary on scientific, historical aspects. Fascinating glimpse of medieval mind. Illustrated. 296pp. 5⅜ × 8¼. (Available in U.S. only) 24609-4 Pa. $5.95

FRANK LLOYD WRIGHT: ARCHITECTURE AND NATURE With 160 Illustrations, Donald Hoffmann. Profusely illustrated study of influence of nature—especially prairie—on Wright's designs for Fallingwater, Robie House, Guggenheim Museum, other masterpieces. 96pp. 9¼ × 10¾. 25098-9 Pa. $7.95

FRANK LLOYD WRIGHT'S FALLINGWATER, Donald Hoffmann. Wright's famous waterfall house: planning and construction of organic idea. History of site, owners, Wright's personal involvement. Photographs of various stages of building. Preface by Edgar Kaufmann, Jr. 100 illustrations. 112pp. 9¼ × 10.
23671-4 Pa. $7.95

YEARS WITH FRANK LLOYD WRIGHT: Apprentice to Genius, Edgar Tafel. Insightful memoir by a former apprentice presents a revealing portrait of Wright the man, the inspired teacher, the greatest American architect. 372 black-and-white illustrations. Preface. Index. vi + 228pp. 8¼ × 11. 24801-1 Pa. $9.95

THE STORY OF KING ARTHUR AND HIS KNIGHTS, Howard Pyle. Enchanting version of King Arthur fable has delighted generations with imaginative narratives of exciting adventures and unforgettable illustrations by the author. 41 illustrations. xviii + 313pp. 6⅛ × 9¼. 21445-1 Pa. $6.50

THE GODS OF THE EGYPTIANS, E. A. Wallis Budge. Thorough coverage of numerous gods of ancient Egypt by foremost Egyptologist. Information on evolution of cults, rites and gods; the cult of Osiris; the Book of the Dead and its rites; the sacred animals and birds; Heaven and Hell; and more. 956pp. 6⅛ × 9¼.
22055-9, 22056-7 Pa., Two-vol. set $21.90

A THEOLOGICO-POLITICAL TREATISE, Benedict Spinoza. Also contains unfinished *Political Treatise*. Great classic on religious liberty, theory of government on common consent. R. Elwes translation. Total of 421pp. 5⅜ × 8½.
20249-6 Pa. $6.95

INCIDENTS OF TRAVEL IN CENTRAL AMERICA, CHIAPAS, AND YUCATAN, John L. Stephens. Almost single-handed discovery of Maya culture; exploration of ruined cities, monuments, temples; customs of Indians. 115 drawings. 892pp. 5⅜ × 8½. 22404-X, 22405-8 Pa., Two-vol. set $15.90

LOS CAPRICHOS, Francisco Goya. 80 plates of wild, grotesque monsters and caricatures. Prado manuscript included. 183pp. 6⅝ × 9⅜. 22384-1 Pa. $4.95

AUTOBIOGRAPHY: The Story of My Experiments with Truth, Mohandas K. Gandhi. Not hagiography, but Gandhi in his own words. Boyhood, legal studies, purification, the growth of the Satyagraha (nonviolent protest) movement. Critical, inspiring work of the man who freed India. 480pp. 5⅜ × 8½. (Available in U.S. only)
24593-4 Pa. $6.95

ILLUSTRATED DICTIONARY OF HISTORIC ARCHITECTURE, edited by Cyril M. Harris. Extraordinary compendium of clear, concise definitions for over 5,000 important architectural terms complemented by over 2,000 line drawings. Covers full spectrum of architecture from ancient ruins to 20th-century Modernism. Preface. 592pp. 7½ × 9⅜. 24444-X Pa. $15.95

THE NIGHT BEFORE CHRISTMAS, Clement Moore. Full text, and woodcuts from original 1848 book. Also critical, historical material. 19 illustrations. 40pp. 4⅝ × 6. 22797-9 Pa. $2.50

THE LESSON OF JAPANESE ARCHITECTURE: 165 Photographs, Jiro Harada. Memorable gallery of 165 photographs taken in the 1930's of exquisite Japanese homes of the well-to-do and historic buildings. 13 line diagrams. 192pp. 8⅜ × 11¼. 24778-3 Pa. $8.95

THE AUTOBIOGRAPHY OF CHARLES DARWIN AND SELECTED LETTERS, edited by Francis Darwin. The fascinating life of eccentric genius composed of an intimate memoir by Darwin (intended for his children); commentary by his son, Francis; hundreds of fragments from notebooks, journals, papers; and letters to and from Lyell, Hooker, Huxley, Wallace and Henslow. xi + 365pp. 5⅜ × 8. 20479-0 Pa. $6.95

WONDERS OF THE SKY: Observing Rainbows, Comets, Eclipses, the Stars and Other Phenomena, Fred Schaaf. Charming, easy-to-read poetic guide to all manner of celestial events visible to the naked eye. Mock suns, glories, Belt of Venus, more. Illustrated. 299pp. 5¼ × 8¼. 24402-4 Pa. $7.95

BURNHAM'S CELESTIAL HANDBOOK, Robert Burnham, Jr. Thorough guide to the stars beyond our solar system. Exhaustive treatment. Alphabetical by constellation: Andromeda to Cetus in Vol. 1; Chamaeleon to Orion in Vol. 2; and Pavo to Vulpecula in Vol. 3. Hundreds of illustrations. Index in Vol. 3. 2,000pp. 6⅛ × 9¼. 23567-X, 23568-8, 23673-0 Pa., Three-vol. set $38.85

STAR NAMES: Their Lore and Meaning, Richard Hinckley Allen. Fascinating history of names various cultures have given to constellations and literary and folkloristic uses that have been made of stars. Indexes to subjects. Arabic and Greek names. Biblical references. Bibliography. 563pp. 5⅜ × 8½. 21079-0 Pa. $7.95

THIRTY YEARS THAT SHOOK PHYSICS: The Story of Quantum Theory, George Gamow. Lucid, accessible introduction to influential theory of energy and matter. Careful explanations of Dirac's anti-particles, Bohr's model of the atom, much more. 12 plates. Numerous drawings. 240pp. 5⅜ × 8½. 24895-X Pa. $5.95

CHINESE DOMESTIC FURNITURE IN PHOTOGRAPHS AND MEASURED DRAWINGS, Gustav Ecke. A rare volume, now affordably priced for antique collectors, furniture buffs and art historians. Detailed review of styles ranging from early Shang to late Ming. Unabridged republication. 161 black-and-white drawings, photos. Total of 224pp. 8⅜ × 11¼. (Available in U.S. only) 25171-3 Pa. $12.95

VINCENT VAN GOGH: A Biography, Julius Meier-Graefe. Dynamic, penetrating study of artist's life, relationship with brother, Theo, painting techniques, travels, more. Readable, engrossing. 160pp. 5⅜ × 8½. (Available in U.S. only) 25253-1 Pa. $3.95

HOW TO WRITE, Gertrude Stein. Gertrude Stein claimed anyone could understand her unconventional writing—here are clues to help. Fascinating improvisations, language experiments, explanations illuminate Stein's craft and the art of writing. Total of 414pp. 4⅝ × 6⅜. 23144-5 Pa. $5.95

ADVENTURES AT SEA IN THE GREAT AGE OF SAIL: Five Firsthand Narratives, edited by Elliot Snow. Rare true accounts of exploration, whaling, shipwreck, fierce natives, trade, shipboard life, more. 33 illustrations. Introduction. 353pp. 5⅜ × 8½. 25177-2 Pa. $7.95

THE HERBAL OR GENERAL HISTORY OF PLANTS, John Gerard. Classic descriptions of about 2,850 plants—with over 2,700 illustrations—includes Latin and English names, physical descriptions, varieties, time and place of growth, more. 2,706 illustrations. xlv + 1,678pp. 8½ × 12¼. 23147-X Cloth. $75.00

DOROTHY AND THE WIZARD IN OZ, L. Frank Baum. Dorothy and the Wizard visit the center of the Earth, where people are vegetables, glass houses grow and Oz characters reappear. Classic sequel to *Wizard of Oz.* 256pp. 5⅜ × 8.
24714-7 Pa. $4.95

SONGS OF EXPERIENCE: Facsimile Reproduction with 26 Plates in Full Color, William Blake. This facsimile of Blake's original "Illuminated Book" reproduces 26 full-color plates from a rare 1826 edition. Includes "The Tyger," "London," "Holy Thursday," and other immortal poems. 26 color plates. Printed text of poems. 48pp. 5¼ × 7. 24636-1 Pa. $3.50

SONGS OF INNOCENCE, William Blake. The first and most popular of Blake's famous "Illuminated Books," in a facsimile edition reproducing all 31 brightly colored plates. Additional printed text of each poem. 64pp. 5¼ × 7.
22764-2 Pa. $3.50

PRECIOUS STONES, Max Bauer. Classic, thorough study of diamonds, rubies, emeralds, garnets, etc.: physical character, occurrence, properties, use, similar topics. 20 plates, 8 in color. 94 figures. 659pp. 6⅛ × 9¼.
21910-0, 21911-9 Pa., Two-vol. set $15.90

ENCYCLOPEDIA OF VICTORIAN NEEDLEWORK, S. F. A. Caulfeild and Blanche Saward. Full, precise descriptions of stitches, techniques for dozens of needlecrafts—most exhaustive reference of its kind. Over 800 figures. Total of 679pp. 8½ × 11. Two volumes. Vol. 1 22800-2 Pa. $11.95
Vol. 2 22801-0 Pa. $11.95

THE MARVELOUS LAND OF OZ, L. Frank Baum. Second Oz book, the Scarecrow and Tin Woodman are back with hero named Tip, Oz magic. 136 illustrations. 287pp. 5⅜ × 8½. 20692-0 Pa. $5.95

WILD FOWL DECOYS, Joel Barber. Basic book on the subject, by foremost authority and collector. Reveals history of decoy making and rigging, place in American culture, different kinds of decoys, how to make them, and how to use them. 140 plates. 156pp. 7⅞ × 10¾. 20011-6 Pa. $8.95

HISTORY OF LACE, Mrs. Bury Palliser. Definitive, profusely illustrated chronicle of lace from earliest times to late 19th century. Laces of Italy, Greece, England, France, Belgium, etc. Landmark of needlework scholarship. 266 illustrations. 672pp. 6⅛ × 9¼. 24742-2 Pa. $14.95

ILLUSTRATED GUIDE TO SHAKER FURNITURE, Robert Meader. All furniture and appurtenances, with much on unknown local styles. 235 photos. 146pp. 9 × 12.
22819-3 Pa. $7.95

WHALE SHIPS AND WHALING: A Pictorial Survey, George Francis Dow. Over 200 vintage engravings, drawings, photographs of barks, brigs, cutters, other vessels. Also harpoons, lances, whaling guns, many other artifacts. Comprehensive text by foremost authority. 207 black-and-white illustrations. 288pp. 6 × 9.
24808-9 Pa. $8.95

THE BERTRAMS, Anthony Trollope. Powerful portrayal of blind self-will and thwarted ambition includes one of Trollope's most heartrending love stories. 497pp. 5⅜ × 8½.
25119-5 Pa. $9.95

ADVENTURES WITH A HAND LENS, Richard Headstrom. Clearly written guide to observing and studying flowers and grasses, fish scales, moth and insect wings, egg cases, buds, feathers, seeds, leaf scars, moss, molds, ferns, common crystals, etc.—all with an ordinary, inexpensive magnifying glass. 209 exact line drawings aid in your discoveries. 220pp. 5⅜ × 8½.
23330-8 Pa. $4.95

RODIN ON ART AND ARTISTS, Auguste Rodin. Great sculptor's candid, wide-ranging comments on meaning of art; great artists; relation of sculpture to poetry, painting, music; philosophy of life, more. 76 superb black-and-white illustrations of Rodin's sculpture, drawings and prints. 119pp. 8⅜ × 11¼.
24487-3 Pa. $6.95

FIFTY CLASSIC FRENCH FILMS, 1912–1982: A Pictorial Record, Anthony Slide. Memorable stills from Grand Illusion, Beauty and the Beast, Hiroshima, Mon Amour, many more. Credits, plot synopses, reviews, etc. 160pp. 8¼ × 11.
25256-6 Pa. $11.95

THE PRINCIPLES OF PSYCHOLOGY, William James. Famous long course complete, unabridged. Stream of thought, time perception, memory, experimental methods; great work decades ahead of its time. 94 figures. 1,391pp. 5⅜ × 8½.
20381-6, 20382-4 Pa., Two-vol. set $23.90

BODIES IN A BOOKSHOP, R. T. Campbell. Challenging mystery of blackmail and murder with ingenious plot and superbly drawn characters. In the best tradition of British suspense fiction. 192pp. 5⅜ × 8½.
24720-1 Pa. $3.95

CALLAS: PORTRAIT OF A PRIMA DONNA, George Jellinek. Renowned commentator on the musical scene chronicles incredible career and life of the most controversial, fascinating, influential operatic personality of our time. 64 black-and-white photographs. 416pp. 5⅜ × 8¼.
25047-4 Pa. $8.95

GEOMETRY, RELATIVITY AND THE FOURTH DIMENSION, Rudolph Rucker. Exposition of fourth dimension, concepts of relativity as Flatland characters continue adventures. Popular, easily followed yet accurate, profound. 141 illustrations. 133pp. 5⅜ × 8½.
23400-2 Pa. $3.95

HOUSEHOLD STORIES BY THE BROTHERS GRIMM, with pictures by Walter Crane. 53 classic stories—Rumpelstiltskin, Rapunzel, Hansel and Gretel, the Fisherman and his Wife, Snow White, Tom Thumb, Sleeping Beauty, Cinderella, and so much more—lavishly illustrated with original 19th century drawings. 114 illustrations. x + 269pp. 5⅜ × 8½.
21080-4 Pa. $4.95

SUNDIALS, Albert Waugh. Far and away the best, most thorough coverage of ideas, mathematics concerned, types, construction, adjusting anywhere. Over 100 illustrations. 230pp. 5⅜ × 8½. 22947-5 Pa. $4.95

PICTURE HISTORY OF THE NORMANDIE: With 190 Illustrations, Frank O. Braynard. Full story of legendary French ocean liner: Art Deco interiors, design innovations, furnishings, celebrities, maiden voyage, tragic fire, much more. Extensive text. 144pp. 8⅜ × 11¾. 25257-4 Pa. $9.95

THE FIRST AMERICAN COOKBOOK: A Facsimile of "American Cookery," 1796, Amelia Simmons. Facsimile of the first American-written cookbook published in the United States contains authentic recipes for colonial favorites—pumpkin pudding, winter squash pudding, spruce beer, Indian slapjacks, and more. Introductory Essay and Glossary of colonial cooking terms. 80pp. 5⅜ × 8½. 24710-4 Pa. $3.50

101 PUZZLES IN THOUGHT AND LOGIC, C. R. Wylie, Jr. Solve murders and robberies, find out which fishermen are liars, how a blind man could possibly identify a color—purely by your own reasoning! 107pp. 5⅜ × 8½. 20367-0 Pa. $2.50

THE BOOK OF WORLD-FAMOUS MUSIC—CLASSICAL, POPULAR AND FOLK, James J. Fuld. Revised and enlarged republication of landmark work in musico-bibliography. Full information about nearly 1,000 songs and compositions including first lines of music and lyrics. New supplement. Index. 800pp. 5⅜ × 8¼. 24857-7 Pa. $14.95

ANTHROPOLOGY AND MODERN LIFE, Franz Boas. Great anthropologist's classic treatise on race and culture. Introduction by Ruth Bunzel. Only inexpensive paperback edition. 255pp. 5⅜ × 8½. 25245-0 Pa. $5.95

THE TALE OF PETER RABBIT, Beatrix Potter. The inimitable Peter's terrifying adventure in Mr. McGregor's garden, with all 27 wonderful, full-color Potter illustrations. 55pp. 4¼ × 5½. (Available in U.S. only) 22827-4 Pa. $1.75

THREE PROPHETIC SCIENCE FICTION NOVELS, H. G. Wells. *When the Sleeper Wakes, A Story of the Days to Come* and *The Time Machine* (full version). 335pp. 5⅜ × 8½. (Available in U.S. only) 20605-X Pa. $6.95

APICIUS COOKERY AND DINING IN IMPERIAL ROME, edited and translated by Joseph Dommers Vehling. Oldest known cookbook in existence offers readers a clear picture of what foods Romans ate, how they prepared them, etc. 49 illustrations. 301pp. 6⅛ × 9¼. 23563-7 Pa. $7.95

SHAKESPEARE LEXICON AND QUOTATION DICTIONARY, Alexander Schmidt. Full definitions, locations, shades of meaning of every word in plays and poems. More than 50,000 exact quotations. 1,485pp. 6½ × 9¼. 22726-X, 22727-8 Pa., Two-vol. set $29.90

THE WORLD'S GREAT SPEECHES, edited by Lewis Copeland and Lawrence W. Lamm. Vast collection of 278 speeches from Greeks to 1970. Powerful and effective models; unique look at history. 842pp. 5⅜ × 8½. 20468-5 Pa. $11.95

THE BLUE FAIRY BOOK, Andrew Lang. The first, most famous collection, with many familiar tales: Little Red Riding Hood, Aladdin and the Wonderful Lamp, Puss in Boots, Sleeping Beauty, Hansel and Gretel, Rumpelstiltskin; 37 in all. 138 illustrations. 390pp. 5⅜ × 8½. 21437-0 Pa. $6.95

THE STORY OF THE CHAMPIONS OF THE ROUND TABLE, Howard Pyle. Sir Launcelot, Sir Tristram and Sir Percival in spirited adventures of love and triumph retold in Pyle's inimitable style. 50 drawings, 31 full-page. xviii + 329pp. 6½ × 9¼. 21883-X Pa. $6.95

AUDUBON AND HIS JOURNALS, Maria Audubon. Unmatched two-volume portrait of the great artist, naturalist and author contains his journals, an excellent biography by his granddaughter, expert annotations by the noted ornithologist, Dr. Elliott Coues, and 37 superb illustrations. Total of 1,200pp. 5⅜ × 8.
Vol. I 25143-8 Pa. $8.95
Vol. II 25144-6 Pa. $8.95

GREAT DINOSAUR HUNTERS AND THEIR DISCOVERIES, Edwin H. Colbert. Fascinating, lavishly illustrated chronicle of dinosaur research, 1820's to 1960. Achievements of Cope, Marsh, Brown, Buckland, Mantell, Huxley, many others. 384pp. 5¼ × 8¼. 24701-5 Pa. $7.95

THE TASTEMAKERS, Russell Lynes. Informal, illustrated social history of American taste 1850's–1950's. First popularized categories Highbrow, Lowbrow, Middlebrow. 129 illustrations. New (1979) afterword. 384pp. 6 × 9.
23993-4 Pa. $8.95

DOUBLE CROSS PURPOSES, Ronald A. Knox. A treasure hunt in the Scottish Highlands, an old map, unidentified corpse, surprise discoveries keep reader guessing in this cleverly intricate tale of financial skullduggery. 2 black-and-white maps. 320pp. 5⅜ × 8½. (Available in U.S. only) 25032-6 Pa. $5.95

AUTHENTIC VICTORIAN DECORATION AND ORNAMENTATION IN FULL COLOR: 46 Plates from "Studies in Design," Christopher Dresser. Superb full-color lithographs reproduced from rare original portfolio of a major Victorian designer. 48pp. 9¼ × 12¼. 25083-0 Pa. $7.95

PRIMITIVE ART, Franz Boas. Remains the best text ever prepared on subject, thoroughly discussing Indian, African, Asian, Australian, and, especially, Northern American primitive art. Over 950 illustrations show ceramics, masks, totem poles, weapons, textiles, paintings, much more. 376pp. 5⅜ × 8. 20025-6 Pa. $6.95

SIDELIGHTS ON RELATIVITY, Albert Einstein. Unabridged republication of two lectures delivered by the great physicist in 1920–21. *Ether and Relativity* and *Geometry and Experience.* Elegant ideas in non-mathematical form, accessible to intelligent layman. vi + 56pp. 5⅜ × 8½. 24511-X Pa. $2.95

THE WIT AND HUMOR OF OSCAR WILDE, edited by Alvin Redman. More than 1,000 ripostes, paradoxes, wisecracks: Work is the curse of the drinking classes, I can resist everything except temptation, etc. 258pp. 5⅜ × 8½. 20602-5 Pa. $4.50

ADVENTURES WITH A MICROSCOPE, Richard Headstrom. 59 adventures with clothing fibers, protozoa, ferns and lichens, roots and leaves, much more. 142 illustrations. 232pp. 5⅜ × 8½. 23471-1 Pa. $3.95

PLANTS OF THE BIBLE, Harold N. Moldenke and Alma L. Moldenke. Standard reference to all 230 plants mentioned in Scriptures. Latin name, biblical reference, uses, modern identity, much more. Unsurpassed encyclopedic resource for scholars, botanists, nature lovers, students of Bible. Bibliography. Indexes. 123 black-and-white illustrations. 384pp. 6 × 9. 25069-5 Pa. $8.95

FAMOUS AMERICAN WOMEN: A Biographical Dictionary from Colonial Times to the Present, Robert McHenry, ed. From Pocahontas to Rosa Parks, 1,035 distinguished American women documented in separate biographical entries. Accurate, up-to-date data, numerous categories, spans 400 years. Indices. 493pp. 6½ × 9¼. 24523-3 Pa. $9.95

THE FABULOUS INTERIORS OF THE GREAT OCEAN LINERS IN HISTORIC PHOTOGRAPHS, William H. Miller, Jr. Some 200 superb photographs capture exquisite interiors of world's great "floating palaces"—1890's to 1980's: *Titanic, Ile de France, Queen Elizabeth, United States, Europa,* more. Approx. 200 black-and-white photographs. Captions. Text. Introduction. 160pp. 8⅜ × 11¾. 24756-2 Pa. $9.95

THE GREAT LUXURY LINERS, 1927–1954: A Photographic Record, William H. Miller, Jr. Nostalgic tribute to heyday of ocean liners. 186 photos of Ile de France, Normandie, Leviathan, Queen Elizabeth, United States, many others. Interior and exterior views. Introduction. Captions. 160pp. 9 × 12. 24056-8 Pa. $10.95

A NATURAL HISTORY OF THE DUCKS, John Charles Phillips. Great landmark of ornithology offers complete detailed coverage of nearly 200 species and subspecies of ducks: gadwall, sheldrake, merganser, pintail, many more. 74 full-color plates, 102 black-and-white. Bibliography. Total of 1,920pp. 8⅜ × 11¼. 25141-1, 25142-X Cloth. Two-vol. set $100.00

THE SEAWEED HANDBOOK: An Illustrated Guide to Seaweeds from North Carolina to Canada, Thomas F. Lee. Concise reference covers 78 species. Scientific and common names, habitat, distribution, more. Finding keys for easy identification. 224pp. 5⅜ × 8½. 25215-9 Pa. $5.95

THE TEN BOOKS OF ARCHITECTURE: The 1755 Leoni Edition, Leon Battista Alberti. Rare classic helped introduce the glories of ancient architecture to the Renaissance. 68 black-and-white plates. 336pp. 8⅜ × 11¼. 25239-6 Pa. $14.95

MISS MACKENZIE, Anthony Trollope. Minor masterpieces by Victorian master unmasks many truths about life in 19th-century England. First inexpensive edition in years. 392pp. 5⅜ × 8½. 25201-9 Pa. $7.95

THE RIME OF THE ANCIENT MARINER, Gustave Doré, Samuel Taylor Coleridge. Dramatic engravings considered by many to be his greatest work. The terrifying space of the open sea, the storms and whirlpools of an unknown ocean, the ice of Antarctica, more—all rendered in a powerful, chilling manner. Full text. 38 plates. 77pp. 9¼ × 12. 22305-1 Pa. $4.95

THE EXPEDITIONS OF ZEBULON MONTGOMERY PIKE, Zebulon Montgomery Pike. Fascinating first-hand accounts (1805–6) of exploration of Mississippi River, Indian wars, capture by Spanish dragoons, much more. 1,088pp. 5⅜ × 8½. 25254-X, 25255-8 Pa. Two-vol. set $23.90

A CONCISE HISTORY OF PHOTOGRAPHY: Third Revised Edition, Helmut Gernsheim. Best one-volume history—camera obscura, photochemistry, daguerreotypes, evolution of cameras, film, more. Also artistic aspects—landscape, portraits, fine art, etc. 281 black-and-white photographs. 26 in color. 176pp. 8⅜ × 11¼. 25128-4 Pa. $13.95

THE DORÉ BIBLE ILLUSTRATIONS, Gustave Doré. 241 detailed plates from the Bible: the Creation scenes, Adam and Eve, Flood, Babylon, battle sequences, life of Jesus, etc. Each plate is accompanied by the verses from the King James version of the Bible. 241pp. 9 × 12. 23004-X Pa. $8.95

HUGGER-MUGGER IN THE LOUVRE, Elliot Paul. Second Homer Evans mystery-comedy. Theft at the Louvre involves sleuth in hilarious, madcap caper. "A knockout."—Books. 336pp. 5⅜ × 8½. 25185-3 Pa. $5.95

FLATLAND, E. A. Abbott. Intriguing and enormously popular science-fiction classic explores the complexities of trying to survive as a two-dimensional being in a three-dimensional world. Amusingly illustrated by the author. 16 illustrations. 103pp. 5⅜ × 8½. 20001-9 Pa. $2.25

THE HISTORY OF THE LEWIS AND CLARK EXPEDITION, Meriwether Lewis and William Clark, edited by Elliott Coues. Classic edition of Lewis and Clark's day-by-day journals that later became the basis for U.S. claims to Oregon and the West. Accurate and invaluable geographical, botanical, biological, meteorological and anthropological material. Total of 1,508pp. 5⅜ × 8½. 21268-8, 21269-6, 21270-X Pa. Three-vol. set $26.85

LANGUAGE, TRUTH AND LOGIC, Alfred J. Ayer. Famous, clear introduction to Vienna, Cambridge schools of Logical Positivism. Role of philosophy, elimination of metaphysics, nature of analysis, etc. 160pp. 5⅜ × 8½. (Available in U.S. and Canada only) 20010-8 Pa. $2.95

MATHEMATICS FOR THE NONMATHEMATICIAN, Morris Kline. Detailed, college-level treatment of mathematics in cultural and historical context, with numerous exercises. For liberal arts students. Preface. Recommended Reading Lists. Tables. Index. Numerous black-and-white figures. xvi + 641pp. 5⅜ × 8½. 24823-2 Pa. $11.95

28 SCIENCE FICTION STORIES, H. G. Wells. Novels, *Star Begotten* and *Men Like Gods,* plus 26 short stories: "Empire of the Ants," "A Story of the Stone Age," "The Stolen Bacillus," "In the Abyss," etc. 915pp. 5⅜ × 8½. (Available in U.S. only) 20265-8 Cloth. $10.95

HANDBOOK OF PICTORIAL SYMBOLS, Rudolph Modley. 3,250 signs and symbols, many systems in full; official or heavy commercial use. Arranged by subject. Most in Pictorial Archive series. 143pp. 8⅛ × 11. 23357-X Pa. $6.95

INCIDENTS OF TRAVEL IN YUCATAN, John L. Stephens. Classic (1843) exploration of jungles of Yucatan, looking for evidences of Maya civilization. Travel adventures, Mexican and Indian culture, etc. Total of 669pp. 5⅜ × 8½. 20926-1, 20927-X Pa., Two-vol. set $9.90

DEGAS: An Intimate Portrait, Ambroise Vollard. Charming, anecdotal memoir by famous art dealer of one of the greatest 19th-century French painters. 14 black-and-white illustrations. Introduction by Harold L. Van Doren. 96pp. 5⅜ × 8½.
25131-4 Pa. $3.95

PERSONAL NARRATIVE OF A PILGRIMAGE TO ALMANDINAH AND MECCAH, Richard Burton. Great travel classic by remarkably colorful personality. Burton, disguised as a Moroccan, visited sacred shrines of Islam, narrowly escaping death. 47 illustrations. 959pp. 5⅜ × 8½.   21217-3, 21218-1 Pa., Two-vol. set $19.90

PHRASE AND WORD ORIGINS, A. H. Holt. Entertaining, reliable, modern study of more than 1,200 colorful words, phrases, origins and histories. Much unexpected information. 254pp. 5⅜ × 8½.   20758-7 Pa. $5.95

THE RED THUMB MARK, R. Austin Freeman. In this first Dr. Thorndyke case, the great scientific detective draws fascinating conclusions from the nature of a single fingerprint. Exciting story, authentic science. 320pp. 5⅜ × 8½. (Available in U.S. only)   25210-8 Pa. $5.95

AN EGYPTIAN HIEROGLYPHIC DICTIONARY, E. A. Wallis Budge. Monumental work containing about 25,000 words or terms that occur in texts ranging from 3000 B.C. to 600 A.D. Each entry consists of a transliteration of the word, the word in hieroglyphs, and the meaning in English. 1,314pp. 6⅞ × 10.
23615-3, 23616-1 Pa., Two-vol. set $31.90

THE COMPLEAT STRATEGYST: Being a Primer on the Theory of Games of Strategy, J. D. Williams. Highly entertaining classic describes, with many illustrated examples, how to select best strategies in conflict situations. Prefaces. Appendices. xvi + 268pp. 5⅜ × 8½.   25101-2 Pa. $5.95

THE ROAD TO OZ, L. Frank Baum. Dorothy meets the Shaggy Man, little Button-Bright and the Rainbow's beautiful daughter in this delightful trip to the magical Land of Oz. 272pp. 5⅜ × 8.   25208-6 Pa. $4.95

POINT AND LINE TO PLANE, Wassily Kandinsky. Seminal exposition of role of point, line, other elements in non-objective painting. Essential to understanding 20th-century art. 127 illustrations. 192pp. 6½ × 9¼.   23808-3 Pa. $4.95

LADY ANNA, Anthony Trollope. Moving chronicle of Countess Lovel's bitter struggle to win for herself and daughter Anna their rightful rank and fortune—perhaps at cost of sanity itself. 384pp. 5⅜ × 8½.   24669-8 Pa. $8.95

EGYPTIAN MAGIC, E. A. Wallis Budge. Sums up all that is known about magic in Ancient Egypt: the role of magic in controlling the gods, powerful amulets that warded off evil spirits, scarabs of immortality, use of wax images, formulas and spells, the secret name, much more. 253pp. 5⅜ × 8½.   22681-6 Pa. $4.50

THE DANCE OF SIVA, Ananda Coomaraswamy. Preeminent authority unfolds the vast metaphysic of India: the revelation of her art, conception of the universe, social organization, etc. 27 reproductions of art masterpieces. 192pp. 5⅜ × 8½.
24817-8 Pa. $5.95

CHRISTMAS CUSTOMS AND TRADITIONS, Clement A. Miles. Origin, evolution, significance of religious, secular practices. Caroling, gifts, yule logs, much more. Full, scholarly yet fascinating; non-sectarian. 400pp. 5⅜ × 8½.
23354-5 Pa. $6.50

THE HUMAN FIGURE IN MOTION, Eadweard Muybridge. More than 4,500 stopped-action photos, in action series, showing undraped men, women, children jumping, lying down, throwing, sitting, wrestling, carrying, etc. 390pp. 7⅞ × 10⅝.
20204-6 Cloth. $21.95

THE MAN WHO WAS THURSDAY, Gilbert Keith Chesterton. Witty, fast-paced novel about a club of anarchists in turn-of-the-century London. Brilliant social, religious, philosophical speculations. 128pp. 5⅜ × 8½.
25121-7 Pa. $3.95

A CEZANNE SKETCHBOOK: Figures, Portraits, Landscapes and Still Lifes, Paul Cezanne. Great artist experiments with tonal effects, light, mass, other qualities in over 100 drawings. A revealing view of developing master painter, precursor of Cubism. 102 black-and-white illustrations. 144pp. 8¾ × 6⅝.
24790-2 Pa. $5.95

AN ENCYCLOPEDIA OF BATTLES: Accounts of Over 1,560 Battles from 1479 B.C. to the Present, David Eggenberger. Presents essential details of every major battle in recorded history, from the first battle of Megiddo in 1479 B.C. to Grenada in 1984. List of Battle Maps. New Appendix covering the years 1967–1984. Index. 99 illustrations. 544pp. 6½ × 9¼.
24913-1 Pa. $14.95

AN ETYMOLOGICAL DICTIONARY OF MODERN ENGLISH, Ernest Weekley. Richest, fullest work, by foremost British lexicographer. Detailed word histories. Inexhaustible. Total of 856pp. 6½ × 9¼.
21873-2, 21874-0 Pa., Two-vol. set $17.00

WEBSTER'S AMERICAN MILITARY BIOGRAPHIES, edited by Robert McHenry. Over 1,000 figures who shaped 3 centuries of American military history. Detailed biographies of Nathan Hale, Douglas MacArthur, Mary Hallaren, others. Chronologies of engagements, more. Introduction. Addenda. 1,033 entries in alphabetical order. xi + 548pp. 6½ × 9¼. (Available in U.S. only)
24758-9 Pa. $11.95

LIFE IN ANCIENT EGYPT, Adolf Erman. Detailed older account, with much not in more recent books: domestic life, religion, magic, medicine, commerce, and whatever else needed for complete picture. Many illustrations. 597pp. 5⅜ × 8½.
22632-8 Pa. $8.95

HISTORIC COSTUME IN PICTURES, Braun & Schneider. Over 1,450 costumed figures shown, covering a wide variety of peoples: kings, emperors, nobles, priests, servants, soldiers, scholars, townsfolk, peasants, merchants, courtiers, cavaliers, and more. 256pp. 8⅜ × 11¼.
23150-X Pa. $8.95

THE NOTEBOOKS OF LEONARDO DA VINCI, edited by J. P. Richter. Extracts from manuscripts reveal great genius; on painting, sculpture, anatomy, sciences, geography, etc. Both Italian and English. 186 ms. pages reproduced, plus 500 additional drawings, including studies for *Last Supper, Sforza* monument, etc. 860pp. 7⅞ × 10¾. (Available in U.S. only) 22572-0, 22573-9 Pa., Two-vol. set $29.90

THE ART NOUVEAU STYLE BOOK OF ALPHONSE MUCHA: All 72 Plates from "Documents Decoratifs" in Original Color, Alphonse Mucha. Rare copyright-free design portfolio by high priest of Art Nouveau. Jewelry, wallpaper, stained glass, furniture, figure studies, plant and animal motifs, etc. Only complete one-volume edition. 80pp. 9⅜ × 12¼. 24044-4 Pa. $8.95

ANIMALS: 1,419 COPYRIGHT-FREE ILLUSTRATIONS OF MAMMALS, BIRDS, FISH, INSECTS, ETC., edited by Jim Harter. Clear wood engravings present, in extremely lifelike poses, over 1,000 species of animals. One of the most extensive pictorial sourcebooks of its kind. Captions. Index. 284pp. 9 × 12.
23766-4 Pa. $9.95

OBELISTS FLY HIGH, C. Daly King. Masterpiece of American detective fiction, long out of print, involves murder on a 1935 transcontinental flight—"a very thrilling story"—NY Times. Unabridged and unaltered republication of the edition published by William Collins Sons & Co. Ltd., London, 1935. 288pp. 5⅜ × 8½. (Available in U.S. only) 25036-9 Pa. $4.95

VICTORIAN AND EDWARDIAN FASHION: A Photographic Survey, Alison Gernsheim. First fashion history completely illustrated by contemporary photographs. Full text plus 235 photos, 1840–1914, in which many celebrities appear. 240pp. 6½ × 9¼. 24205-6 Pa. $6.95

THE ART OF THE FRENCH ILLUSTRATED BOOK, 1700–1914, Gordon N. Ray. Over 630 superb book illustrations by Fragonard, Delacroix, Daumier, Doré, Grandville, Manet, Mucha, Steinlen, Toulouse-Lautrec and many others. Preface. Introduction. 633 halftones. Indices of artists, authors & titles, binders and provenances. Appendices. Bibliography. 608pp. 8⅜ × 11¼. 25086-5 Pa. $24.95

THE WONDERFUL WIZARD OF OZ, L. Frank Baum. Facsimile in full color of America's finest children's classic. 143 illustrations by W. W. Denslow. 267pp. 5⅜ × 8½. 20691-2 Pa. $5.95

FRONTIERS OF MODERN PHYSICS: New Perspectives on Cosmology, Relativity, Black Holes and Extraterrestrial Intelligence, Tony Rothman, et al. For the intelligent layman. Subjects include: cosmological models of the universe; black holes; the neutrino; the search for extraterrestrial intelligence. Introduction. 46 black-and-white illustrations. 192pp. 5⅜ × 8½. 24587-X Pa. $6.95

THE FRIENDLY STARS, Martha Evans Martin & Donald Howard Menzel. Classic text marshalls the stars together in an engaging, non-technical survey, presenting them as sources of beauty in night sky. 23 illustrations. Foreword. 2 star charts. Index. 147pp. 5⅜ × 8½. 21099-5 Pa. $3.50

FADS AND FALLACIES IN THE NAME OF SCIENCE, Martin Gardner. Fair, witty appraisal of cranks, quacks, and quackeries of science and pseudoscience: hollow earth, Velikovsky, orgone energy, Dianetics, flying saucers, Bridey Murphy, food and medical fads, etc. Revised, expanded In the Name of Science. "A very able and even-tempered presentation."—The New Yorker. 363pp. 5⅜ × 8.
20394-8 Pa. $6.50

ANCIENT EGYPT: ITS CULTURE AND HISTORY, J. E Manchip White. From pre-dynastics through Ptolemies: society, history, political structure, religion, daily life, literature, cultural heritage. 48 plates. 217pp. 5⅜ × 8½. 22548-8 Pa. $5.95

SIR HARRY HOTSPUR OF HUMBLETHWAITE, Anthony Trollope. Incisive, unconventional psychological study of a conflict between a wealthy baronet, his idealistic daughter, and their scapegrace cousin. The 1870 novel in its first inexpensive edition in years. 250pp. 5⅜ × 8½. 24953-0 Pa. $5.95

LASERS AND HOLOGRAPHY, Winston E. Kock. Sound introduction to burgeoning field, expanded (1981) for second edition. Wave patterns, coherence, lasers, diffraction, zone plates, properties of holograms, recent advances. 84 illustrations. 160pp. 5⅜ × 8¼. (Except in United Kingdom) 24041-X Pa. $3.50

INTRODUCTION TO ARTIFICIAL INTELLIGENCE: SECOND, EN-LARGED EDITION, Philip C. Jackson, Jr. Comprehensive survey of artificial intelligence—the study of how machines (computers) can be made to act intelligently. Includes introductory and advanced material. Extensive notes updating the main text. 132 black-and-white illustrations. 512pp. 5⅜ × 8½. 24864-X Pa. $8.95

HISTORY OF INDIAN AND INDONESIAN ART, Ananda K. Coomaraswamy. Over 400 illustrations illuminate classic study of Indian art from earliest Harappa finds to early 20th century. Provides philosophical, religious and social insights. 304pp. 6⅜ × 9⅜. 25005-9 Pa. $8.95

THE GOLEM, Gustav Meyrink. Most famous supernatural novel in modern European literature, set in Ghetto of Old Prague around 1890. Compelling story of mystical experiences, strange transformations, profound terror. 13 black-and-white illustrations. 224pp. 5⅜ × 8½. (Available in U.S. only) 25025-3 Pa. $6.95

ARMADALE, Wilkie Collins. Third great mystery novel by the author of *The Woman in White* and *The Moonstone*. Original magazine version with 40 illustrations. 597pp. 5⅜ × 8½. 23429-0 Pa. $9.95

PICTORIAL ENCYCLOPEDIA OF HISTORIC ARCHITECTURAL PLANS, DETAILS AND ELEMENTS: With 1,880 Line Drawings of Arches, Domes, Doorways, Facades, Gables, Windows, etc., John Theodore Haneman. Sourcebook of inspiration for architects, designers, others. Bibliography. Captions. 141pp. 9 × 12. 24605-1 Pa. $6.95

BENCHLEY LOST AND FOUND, Robert Benchley. Finest humor from early 30's, about pet peeves, child psychologists, post office and others. Mostly unavailable elsewhere. 73 illustrations by Peter Arno and others. 183pp. 5⅜ × 8½. 22410-4 Pa. $3.95

ERTÉ GRAPHICS, Erté. Collection of striking color graphics: *Seasons, Alphabet, Numerals, Aces* and *Precious Stones*. 50 plates, including 4 on covers. 48pp. 9⅜ × 12¼. 23580-7 Pa. $6.95

THE JOURNAL OF HENRY D. THOREAU, edited by Bradford Torrey, F. H. Allen. Complete reprinting of 14 volumes, 1837–61, over two million words; the sourcebooks for *Walden*, etc. Definitive. All original sketches, plus 75 photographs. 1,804pp. 8½ × 12¼. 20312-3, 20313-1 Cloth., Two-vol. set $80.00

CASTLES: THEIR CONSTRUCTION AND HISTORY, Sidney Toy. Traces castle development from ancient roots. Nearly 200 photographs and drawings illustrate moats, keeps, baileys, many other features. Caernarvon, Dover Castles, Hadrian's Wall, Tower of London, dozens more. 256pp. 5⅜ × 8¼.

24898-4 Pa. $5.95

AMERICAN CLIPPER SHIPS: 1833–1858, Octavius T. Howe & Frederick C. Matthews. Fully-illustrated, encyclopedic review of 352 clipper ships from the period of America's greatest maritime supremacy. Introduction. 109 halftones. 5 black-and-white line illustrations. Index. Total of 928pp. 5⅜ × 8½.
25115-2, 25116-0 Pa., Two-vol. set $17.90

TOWARDS A NEW ARCHITECTURE, Le Corbusier. Pioneering manifesto by great architect, near legendary founder of "International School." Technical and aesthetic theories, views on industry, economics, relation of form to function, "mass-production spirit," much more. Profusely illustrated. Unabridged translation of 13th French edition. Introduction by Frederick Etchells. 320pp. 6⅛ × 9¼. (Available in U.S. only)
25023-7 Pa. $8.95

THE BOOK OF KELLS, edited by Blanche Cirker. Inexpensive collection of 32 full-color, full-page plates from the greatest illuminated manuscript of the Middle Ages, painstakingly reproduced from rare facsimile edition. Publisher's Note. Captions. 32pp. 9⅜ × 12¼.
24345-1 Pa. $4.95

BEST SCIENCE FICTION STORIES OF H. G. WELLS, H. G. Wells. Full novel *The Invisible Man*, plus 17 short stories: "The Crystal Egg," "Aepyornis Island," "The Strange Orchid," etc. 303pp. 5⅜ × 8½. (Available in U.S. only)
21531-8 Pa. $6.95

AMERICAN SAILING SHIPS: Their Plans and History, Charles G. Davis. Photos, construction details of schooners, frigates, clippers, other sailcraft of 18th to early 20th centuries—plus entertaining discourse on design, rigging, nautical lore, much more. 137 black-and-white illustrations. 240pp. 6⅛ × 9¼.
24658-2 Pa. $6.95

ENTERTAINING MATHEMATICAL PUZZLES, Martin Gardner. Selection of author's favorite conundrums involving arithmetic, money, speed, etc., with lively commentary. Complete solutions. 112pp. 5⅜ × 8½.
25211-6 Pa. $2.95

THE WILL TO BELIEVE, HUMAN IMMORTALITY, William James. Two books bound together. Effect of irrational on logical, and arguments for human immortality. 402pp. 5⅜ × 8½.
20291-7 Pa. $7.50

THE HAUNTED MONASTERY and THE CHINESE MAZE MURDERS, Robert Van Gulik. 2 full novels by Van Gulik continue adventures of Judge Dee and his companions. An evil Taoist monastery, seemingly supernatural events; overgrown topiary maze that hides strange crimes. Set in 7th-century China. 27 illustrations. 328pp. 5⅜ × 8½.
23502-5 Pa. $5.95

CELEBRATED CASES OF JUDGE DEE (DEE GOONG AN), translated by Robert Van Gulik. Authentic 18th-century Chinese detective novel; Dee and associates solve three interlocked cases. Led to Van Gulik's own stories with same characters. Extensive introduction. 9 illustrations. 237pp. 5⅜ × 8½.
23337-5 Pa. $4.95

Prices subject to change without notice.
Available at your book dealer or write for free catalog to Dept. GI, Dover Publications, Inc., 31 East 2nd St., Mineola, N.Y. 11501. Dover publishes more than 175 books each year on science, elementary and advanced mathematics, biology, music, art, literary history, social sciences and other areas.